PATHWAY TOWARDS COLLEGE MATHEMATICS

By James Fulton
&
Elizabeth Chu

Suffolk County Community College

Mathematics Department
Suffolk County Community College
533 College Road
Selden, NY 11784

Preface to the Student:

Mathematics is a language by which we try and make sense out of the world around us. As a student you have probably studied mathematics for the better part of the last 13 years, if you just came out of high school. Since you are taking this course many of the fundamental ideas and concepts you either did not retain or never fully understood in the first place. This book and course will serve as the pathway towards the higher mathematics level you need to be successful at the college level.

The two critical mathematics concepts you need to master at this level are that of number and basic ideas in geometry. Without this foundation you will not be able to succeed. Now, while the idea of number may seem trivial, since it is one of the first things a child learns. It is, in reality, quite advanced and abstract. In fact, to fully understand it, most mathematicians don't reach that level until they are in graduate school. Why is it so complicated? It is because the concept of number moves from beyond simply counting whole objects to measuring lengths. Lengths can take on any size in mathematics, which poses some very confusing conundrums and opens up the idea of things called irrational numbers. Numbers that do not exist in our real physical world, but only reside in a specialized mathematical universe. This evokes some deep philosophical discussions.

Although this textbook will not take you down the path of fully understanding all aspects of the number concept, it will, however, start you on the journey. If you are to master the number concept enough to move on to a higher mathematical level, you must first fully understand three important types of numbers. They are what we call the whole numbers, the fractions and decimals, and the negative numbers. Now while whole numbers, also called the counting numbers are the basic building blocks of all numbers, their properties can at times be confusing to some, but need to be fully understood. The idea of fractions, however, is an extremely abstract concept relative to whole numbers. It is a change from counting objects to measuring them. Such a change introduces a major reinterpretation of the number concept, which causes most students to get lost. Even most calculus students today, don't fully understand the fraction concept, and as such, have just adjusted to following rules and standard algorithms which they often mix up. Then after the fraction concept comes the concept of negative numbers. With negative numbers you can not only "measure" things, but you can also measure their relative location as well. This new concepts introduces more rules that further abstracts the number concept.

Once we begin to measure objects, the need to understand attributes about objects becomes important. That is where the concept of geometry comes in. Instead of measuring lengths we need to measure areas and volumes. What is even more confusing, we have different systems and scales used to measure these quantities with. Thus, you will need to master the concept of units, not just for your mathematics courses, but more importantly for your science courses. This just scratches the surface of what is needed to advance beyond this course.

Now, the only way to learn anything is to practice it and use it over-and-over again. You will not be able to watch the instructor and then flawlessly perform the same steps, just like you would not be able to play a piece of music; paint a picture; perform surgery; drive a car; or play a particular sport, by simply watching a person and then repeating what you saw them do. You must try it yourself, again-and-again-and-again …

As you will come to know, mathematics is a very precise language, therefore, we need an element of precision in how we approach it. However, when new mathematics is created or discovered, the process by which this happens is almost always informal, and with a lot of trial and error. When the subject matter is taught, though, the material is presented in the formal context first, and then informal examples are used to highlight the theory after the fact. This is quite contrary to how it was created in the first place. We will aim for more of a mixed, informal-approach first, and then summarize the formal results and definitions at the end of each section, not at the beginning.

Finally, this course is a prerequisite for the next developmental math course before you can take a college level mathematics or science courses, such as intermediate algebra, statistics, and laboratory science courses. You need to become proficient in what is contained in these pages if you hope to succeed in your next math course, which is required before you can advance to the college level. Even though you have studied mathematics for many years, don't take it for granted that you know it, because you will find out you do not. Open up your mind to being taught these fundamental concepts differently, and, MOST IMPORTANTLY, do the homework!

Preface to the Instructor:

This book is intended for a first course in a developmental sequence of mathematics courses for college students that need to achieve a basic number sense, so that they can move on to a course covering some basic algebra and other necessary topics.

The approach is a non-traditional approach where the focus is on understanding concepts upon which standard algorithms of calculation are based. In particular, a large part of the book deals with the concept of fractions, since this has shown to be critical for success in higher mathematics courses. We want the students to grasp the concept of fraction as being similar to whole numbers representing points on a number line. The arithmetic algorithms for fractions are presented for understanding and not just computation. It is only after students truly understand the idea of fractions can they generalize the ideas learned to, decimals, unit conversions, and proportions.

The other common theme throughout the book is the idea of using mathematics, and numbers in particular, for making measurements and calculating results from these measurements. This is why basic geometric ideas and formula calculations are a large part of this book.

Obviously, we will not be able to correct for many years of misunderstanding in a single semester course. Our hope, however, is to give the students enough understanding for them to be successful in their follow-up

Acknowledgments:

We would like to thank the College Administration for giving us the time to pursue this effort and in allowing us to present this new approach to our students. In particular, we'd like to thank President Shaun McKay and Associate Vice President for Academic Affairs Paul Beaudin for all their support and encouragement. Furthermore, the chairs of our departments on the three campuses of this college were extremely supportive. A special thanks to the chairs: Dennis Reissig, Ted Koukounas, and John Jerome for their support.

TABLE OF CONTENTS

Chapter 4: Arithmetic of Fractions

Chapter 5. Decimal Fractions (Decimals)

Chapter 6: Percents

Chapter 11: Solving Equations

INTRODUCTION

Measurement – Measuring Your Way Through Mathematics

Students always ask – 'Why do I have to study math?,' or 'When will I ever use this?' The answer quite frankly is simple. Math is everything! It is everywhere around us, and yes, it is true that we may not need the specific math skill we are learning at a particular time, but if we would only open up ourselves to the possibilities, we might notice that it could be useful in some other setting or other time in the future. For you see, math is really all around us. It truly is the language of the universe. It is a language that allows us to describe all the attributes of this thing we call the universe, in a very definable and precise way. Mathematics is the language of patterns, and patterns are what make up the universe!

The world is full of patterns. The sun "rises" and "sets," the seasons come and go, and cicadas breed in unique and consistent time delays, all forming cyclical patterns. Birds fly in formation, spiders spin webs, trees grow branches, flowers form seed packs and petals, and bees build hives, all forming rich geometric patterns. The weather dramatically changes, and the stock market seems to erratically rise and fall, all giving rise to seemingly chaotic patterns.

When Aristotle sought to define logic, the essence of human thought and reasoning, as a way to make language more precise, little did he foresee that it would be mathematics that would some day add true precision to it, and clear up the ambiguities he sought to dismiss. He never saw the insights of Leibniz, Boole, Frege and Russell that would show that even the idea of human thought could potentially be understood by and reduced to mathematics – the mathematical patterns associated with human reason.

Mathematics has been called the Queen of all the Sciences, essentially because; mathematics is the science of recognizing, analyzing, and deciphering patterns. Patterns make the world the way it is and understanding the code needed to decipher them gives us power to predict and alter their effects. That code is written in the language of mathematics. The only way we can grasp these patterns is to use our senses to observe, characterize, and quantify their unique attributes. We accomplish the important act of quantification through measuring. Our understanding hinges on our ability to reproducibly measure the attributes we observe.

To measure means we determine either the amount, size, or degree of something, by either counting, using a device calibrated with some set of standardized attribute units, or by comparing it to an object with a known attribute value. Measuring is the most important tool of any mathematician. Without it, we would have only vague words to try and describe what we "see," but no concrete reproducible way to determine if the pattern persists in a precise unchanging way. We would have no way to determine if the pattern is caused by a fundamental law of nature, or simply by the appearances of what we think we "see".

In this textbook we will embark on the journey of gaining an understanding of the art of measuring. It is a valuable skill for all to have, not just for mathematicians and scientists, but for anyone to survive in this world. Measuring is how we make sense and navigate our way through this complex world and life.

To be able to master the art of measuring means we need to be proficient with several concepts. The first concept is that of **number**. Now by number we shall mean, whole numbers, fractions, decimals (decimal fractions), irrational numbers, exponential numbers, and even negative numbers. We furthermore need to be aware of the idea of accuracy and precision of our measurements related to these numbers. This leads to the idea of obtaining approximate values for our measurements, and also being able to effectively reproduce our results.

The next important concept related to measuring is **geometry**. Even the name geometry; (from **geo** meaning 'earth', and **metria** meaning 'measure'), or earth–measure, involves the concept of measurement in its name. The shapes of objects and their relationships are important tools we use to characterize objects and things around us. How and what we measure geometrically is very important.

We also need to understand how measuring different attributes of things are related to each other. Thus we need to see how our measured data are related to each other, and so we need to know how to create and interpret relations between measure values. We do this in the form of graphs of data, and this inevitably leads to the idea of **algebra** and its relation to measuring. However, algebra will be a subject for the next course in the sequence, and will only be talked about in a simple introductory context.

This textbook will use the concept of measuring, which we have just stated is critical to our ability to live in this world, to unify several branches of mathematics related to; numbers and geometry. Along with whatever we are introducing or defining, we will also consider; how does this affect how we measure? Furthermore, the idea of measuring is essential to any scientific investigation. Thus, it is central in any laboratory science class, as well as any mathematics or quantification related course. Hence, another purpose of this textbook is to equip students with the necessary skills to succeed in the next course which will prepare you for a college level science course as well as a college level mathematics course.

MEASUREMENT

One of the most important aspects of mathematics has to do with solving problems. Not the concocted word-problems we have all become frustrated by, but problems of real consequence. We frequently need to know "how much?", and "of what?" How much time do we have? How far do we have to travel? How high up is it? How cold is it? How much flour do I add? How much rain did we get? How long does the board need to be? How much fuel do I need? The list can go on.

To be able to answer the **how-much** question, we need to know how to measure. Whether it is with a ruler, a weight-scale, or a thermometer, we need to know about measuring. It is how we accurately compare quantities to see which is larger, or how we determine how much of a certain quantity is present relative to another quantity, or how we properly mix ingredients, or how we determine if things fit together properly or not. It is a vital and necessary skill.

All measurements are relative and are made up of two distinct parts. The first part answers the "how-much?" question. It is the numerical characterization of the measurement, such as 1, 2, 3, 6.5, or 10.75. The second part answers the "of what?" question. It identifies what we are measuring — it is the **units** part of our measurement, which also has two aspects, its type or attribute, and its scale. Any measurement must contain both a number and a unit (attribute), otherwise it is useless and meaningless. When measuring we are really asking "How much", and "of what, are we measuring?" The number expresses the magnitude, and the unit expresses the type of quantity we are considering, but given in a specific scale. See the figure below:

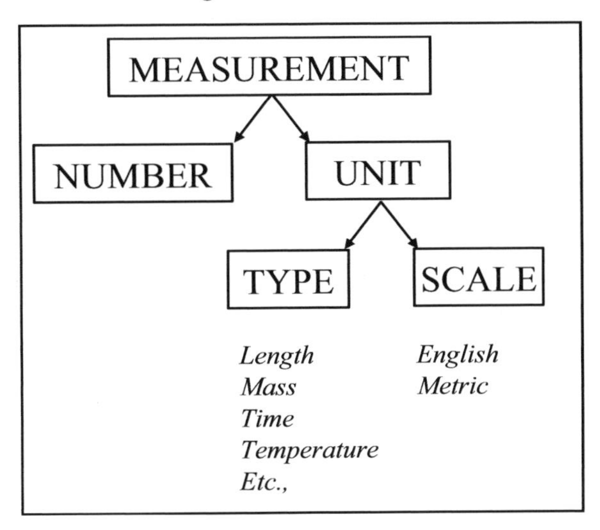

Examples of measurements and units: 25 mm, 16 in^2, 35.7 g, 2.5 gal., 64 km, 153 miles, 36°F, 5 lbs, and 12 oz., 1.3 kg, 75 mph, 16 Amps., 39 cm^3, etc.

In mathematics, we often disregard the units, which does a great disservice to problem solving. Units are key to every problem, so in this textbook we will focus on, and pay close attention to units.

UNITS

Some common attributes or units are:

- Size: length, area, and volume
- Weight
- Distance
- Temperature
- Luminosity (Brightness)
- Electric Potential
- pH level
- Amount (mole)
- Density
- Forces (including pressure)
- Energy
- Velocity
- Time
- Mass

Each of the above quantities is a particular type of unit. All of the above units fall into one of two different categories. It is either a fundamental unit or a derived unit.

Fundamental Units

A fundamental unit (also called a base unit) is a unit that is **dimensionally-independent** from the other units. In other words, they are fundamentally different and cannot be derived from another unit. There are seven fundamental or base units, and they are:

Length
Mass
Time
Temperature
Electric Charge
Amount (mole)
Luminosity

We'll focus on the first four fundamental quantities, as these are more common in our day-to-day living.

Length

Length is the fundamental geometric measure, which is associated with the linear size or location of an object – How long is it? How tall is it? How far away is it?

Mass

Mass measures the amount of matter in an object. Sometimes it is mistakenly referred to as weight. Weight, on the other hand, is the gravitational attraction between the earth and the mass of the object. The object's mass is attracted to the earth, and the earth is attracted to the object, and the mutual strength of this attraction is called the weight of the object. A strange fact is that an object can have mass, yet still be weightless. For example, if we go way out into space we float, as we are weightless, but we still have a "bulkiness" associated with us. This bulkiness is our mass.

Since most of the measurements we come in contact with are done on earth, we won't worry about the distinction. Here on earth, if two things have the same mass, they will also have the same weight. Even though we might treat mass and weight the same, technically we should distinguish between the two. If you took a bowling ball or a car to the moon, its weight would be less, but its mass would remain the same.

Time

Time measures the passing of events. It tells us **how long** something takes to happen.

It is interesting to note that time, like length and mass, is an independent quantity. However, Einstein has so strangely shown us, these quantities are intimately connected. Time can change based upon an object changing its location (length), or if it is in the presence of a very large mass. Very strange results indeed! Something even scientists still do not fully understand!

Temperature

Temperature is a quantity that measures the degree of **hotness** or **coldness** of an environment or a body. The hotness is a measure of how much **vibrational-energy** (kinetic energy) is contained within the molecules of the material in question. The more the atoms vibrate, the higher the temperature.

Molecules that do not move, do not have any measurable vibrational-energy and are considered to have a temperature that is called absolute zero. In practice this never occurs. There really is no absolute zero!

Derived Units

Everything else we measure we obtain from the above fundamental quantities, and these units are called derived units. The fundamental units provide the basis from which the units for all other quantities are derived. Just as the letters in the alphabet give rise to all of the words in the English language. Similarly, the base units create all the other types of units we need and use in our calculations. We obtain these derived units by multiplying and dividing the fundamental units by each other.

EXAMPLES:

Velocity is distance (length) divided by time (length/time), e.g. miles per hour (mph) or meters per second (mps).

Volume is the product of three length measurements ($length^3$), e.g. cubic inches (in^3) or cubic meters (m^3). Note how we use exponents to simplify the representation.

Density is mass divided by volume, where volume is another derived unit. Thus, we can express density in terms of fundamental units as mass divided by length cubed ($mass/length^3$), e.g. grams per centimeter cubed (g/cm^3) or kilograms per meter cubed (kg/m^3).

Examples of other derived quantities:

Area is the product of length times length (width)
Acceleration is length divided by seconds squared
Force is mass times acceleration (another derived unit).
etc.

Any time we measure an attribute, there are two parts to consider. The first is the number or numerical part, and the second is the unit part. The unit part is really two things. The first is the type or attribute we are measuring, and the second is the scale on which we are referencing it to. For now we will not spend a lot of time on the scale part of the unit concept, but will instead focus on the previous exposure you have had to the different scales for length, such as; inches, feet, mile, meter, centimeter, millimeter, and kilometer, as well as for mass, such as; ounce, pound, gram, and kilogram. In Chapter 12 we spend more time introducing the two major scales; the metric and the English, and show in detail how they are connected and how we can translate (convert) between and within both scale systems. However, our primary focus will be on the Metric System, since that is the system essentially used in our scientific study. It is a shame that here in the US we have to learn two different scales. This places another unnecessary burden on our children in a subject that is already quite complicated: and we wonder why our students seem to be behind other nations when it comes to mathematics. We handicap our children and then wonder why they can't compete!

For now, we move on to the first fundamental topic related to measurements: the concept of number. A number as we shall see is more than just a way to count whole objects. Within the concept of measurement there is a need to go beyond simply counting whole objects. However, the concept of number and counting objects is still the same. It is our interpretation of what is an object that we are counting that has to be expanded. Instead of discrete objects, we must extend our counting to portions of a scale, such as in geometry when we must measure how long an object is, i.e. how many centimeters or inches long it may be. Thus, a single object can actually be thought of as being made up of parts of a whole when the idea of size and distance is introduced into our number concept.

In the next few chapters we expand our idea of number, but the same intuitive model of simply counting and multiplying whole objects still applies, and can be used to assist us as we work through some seemingly abstract concepts that have very concrete and practical analogies that can be useful in helping us understand them. To this end, we will often use geometric ideas, concepts, and objects, as we extend our understanding of number. This is an idea we can fall back upon whenever we become confused when we consider the arithmetic of numbers. We will expand upon this idea further in Part 1 of the textbook.

What you should take away from this part of the book, is just to have a very vague understanding of units. You need to know that a unit is an attribute, and there are roughly 4 very important units/attributes we need to be aware of; length, mass (weight), time, and temperature. By multiplying the unit of length we get two very important derived units; area and volume. Thus, you should know that some units for length are; inches, feet, yards, miles, meters, centimeters, millimeters, and kilometers. These are just a few of the important ones. There are others, but these you should have the idea are all used to measure a length. At this point, however, you do not need to know how to translate from one unit scale to another. That we will show later in the book. Right now you just need to know these units measure a length.

You should also know that some units of mass are; pound, ton, gram, kilogram, and milligram are all units to measure the mass or how much "stuff" something has. Again, there are others, but these are the important ones, for now. The units for time are fairly self-explanatory as; seconds, minutes, hours, days, months, years, light-years, etc. The units for temperature are either degrees Fahrenheit (^{o}F), or degrees Celsius (^{o}C).

Finally, we want you to fully understand the two very important derived units; area, and volume. The common units for area are; square inches (in^2), square feet (ft^2), square meters (m^2), square centimeters (cm^2), to name some of the more common units. The common units for volume are; cubic inches (in^3), cubic feet (ft^3), cubic meters (m^3), cubic centimeters (cm^3), to name some of the more common units. If you have this very basic knowledge, that will be fine for now. Later in the book we'll explore the different scales we have introduced, and how to convert from one scale to another. But that won't happen until much later. Right now you just need a good intuitive idea of which units are which and what they are used to measure.

EXERCISES: Attributes & Measurements

1. Come up with a list of at least 10 attributes of an object or an event that you can measure.
2. For the quantities in the table below, list all the units that can be used to measure the given attributes. Choose these units from the list beneath this table.

Different unit names for the same attribute	
Attribute	**Unit names**
length	
area	
volume	
mass (weight)	
temperature	
time	
density	
velocity	

board-feet
centigrade
centigrams
centimeters
centuries
cubic centimeters
cubic yards
cups
days
decades
°Celsius
°Fahrenheit
feet
feet per second
fluid ounces
gallons
grams per cubic centimeter
grams per liter
grams per milliliter
grams
hours
inches
kilograms
kilometers
kilometers per hour
light-years
liters
meters
meters per second
metric tons
micrograms
microns
mils
miles
miles per hour
milligrams
millennia
milliliters
milliseconds
minutes
nanometers
nanoseconds
ounces
pints
pounds per cubic foot
quarts
square inches
square centimeters
tons
yards

3. Given the units from problem 2 above, identify which are fundamental and which are derived.
4. Match the appropriate units from the list provided below, with the descriptions a thru p. Each unit will appear only once and all units will be used.

degrees Celsius	kilometers	miles per hour	pounds
degrees Fahrenheit	kilometers per hour	milligrams	stories
feet	liters	millimeters	tons
gallons	miles	meters	yards

   a. The world's tallest building is the Burj Dubai in the United Arab Emirates and stands 2,717 _________ tall.

b. America's tallest building, the Willis Tower (formerly Sears Tower) in Chicago is 108 ________ high.

c. The human body contains approximately 5.6 ________ of blood.

d. The longest in-air pass was by Don Meredith to Bob Hays of the Dallas Cowboys in 1966 for a total of 83 _______.

e. The hottest temperature ever recorded on Earth was El Azizia in Libya on September 13, 1922 and was 57 ________.

f. The deepest spot in the ocean is the Mariana Trench in the Pacific Ocean and is 11 _____ deep.

g. The highest waterfall in the world is Angel Falls in Venezuela and is 979 _____ high.

h. The coldest temperature ever recorded was -128.6 ____ in Vostok, Antarctica in 1983.

i. The fastest regular wind speed was at Mt. Washington in New Hampshire on April 12th, 1934 with a speed of 372 ________

j. The largest recorded hailstone to ever fall landed in Aurora Nebraska in June of 2003. It had a circumference of 476 _____.

k. The largest seed in the world is from the coc-de-mer coconut tree, which may weigh as much as 30 _____.

l. The world's fastest manned aircraft is the X-15 with a record speed of 4,520_____.

m. Coca-Cola® is currently the most popular soft drink in the World with more than 9.6 Billion _______ consumed each day.

n. The highest spot on land is Mt Everest with a height of 5.5 ________above sea level.

o. A 4 oz. piece of chicken liver contains approximately 641 ______ of cholesterol

p. The heaviest blue whale ever weighed, was over 190 __________.

In the first chapter of the text, we begin with the idea of number. A concept you have seen many times before. It is our hope that if you have had problems in the past grasping some of the concepts related to numbers, that we can help you to master these concepts. For others it will be a review of something you may not have seen for a while. For us, the authors, it is an opportunity to make sure that all students are starting with the same foundation that we can begin to build upon.

CHAPTER 1

Introduction to Whole Numbers

Numbers are the fundamental building blocks on which all of mathematics is formed and based. You have worked with numbers for most of your life, and have studied them and their properties extensively in school. Your previous exposure to the number concept is why that in this chapter we start by providing a quick overview of numbers and their properties, which are essential for a thorough understanding of the mathematical concepts required in many of your technical college courses. After this brief review/overview, we take a step back and examine the number concept all over again, but in greater detail. We realize that for many of you it never really made sense the first time you saw it or, out of lack of interest, you never really paid much attention to it. That is why in sections 1.2 and beyond we take a fresh look at the number concept, and hope you will open your mind and become more inquisitive about this fundamental idea.

Our goal is understanding, not memorization. Take the time to fully understand what is presented, and not just simply look for the quickest way to memorize some vague technique that will give you the correct answer. Getting the correct answer without knowing why is more dangerous than making a careless mistake in your solution, but knowing why you were wrong.

1.1 A FRESH LOOK AT NUMBERS: Understanding Numbers, Place–Value, and more

In your previous courses in high school you have studied the various types of numbers. In this section, however, we wish to revisit the introduction of these numbers, and focus more on the number concept. We realize that you have seen this material many times before, but it is our hope that we can give you some new tools and perspectives to better understand the associated concepts related to the different types of numbers. This will be especially true when we introduce fractions and negative numbers later on. These are two very fundamental and difficult concepts that students need to fully comprehend.

Defining the Number Concept and Axiomatic Systems

Let's begin again. We start with a conceptual definition of number. We should point out that this is not an easy task mathematically. What seems so obvious and simple is much more involved than we can get into at this level. Strangely it brings deep philosophical issues to light that will not be discussed in this book.

A mathematical description of number introduces numbers, along with the operations we define on them, using something called an axiomatic system. This is a formal system based upon principles of logic that allows us to avoid mistakes and paradoxical results as we develop the concepts and their relationships.

An axiomatic system consists of an undefined term/terms, along with statements called axioms that are assumed to be true. The axioms are the rules of our axiomatic game. Definitions, in terms of the undefined term/terms, of objects are then created so that properties of and relationships between these newly defined objects can be proven. A complete mathematical theory of numbers and their properties and relationships is obtained by proving new statements, called theorems, using only the axioms, logic, and previous theorems.

Parts of an Axiomatic System:

1. Undefined terms/primitive terms
2. Defined terms
3. Axioms/postulates-accepted unproved statements
4. Theorems-proved statements

In the theory of numbers, our undefined object is the concept of what we mean by the word number. We will start with pseudo–definition where we instead describe what we'd like our numbers object to do.

Pseudo–Definition: A **number** is a mathematical object used to count, measure, and label.

Now we could start out in a very formal way by listing all the axioms and definitions and begin to prove new theorems, but we will not. We did, however, want to introduce the idea of an axiomatic system, so that you would know there is a logically consistent approach for developing each concept and idea we come across. We will, from time–to–time, refer to this implied connection as we develop the subject further, but for now we'll proceed less formally.
We take a less formal/rigorous approach because the concept of number, believe it or not, is quite hard to define in a logically consistent mathematical way, that allows all the types of numbers we discussed earlier to be considered as numbers. The above definition is not a true definition of number because the concepts of count, measure, and label all require that we already have the concept of number defined to be able to define them. Thus, our definition above is circular. As a result we simply take this as a pseudo–definition and use our intuition about what numbers are, and proceed from there. This is not rigorous, but necessary if we are going to move forward without losing all readers as we proceed.

From our working definition we see that a number can be used to count individual objects, but that it is also used to measure quantities. The quality of being able to measure is what requires "in–between" values for our numbers. We could be asked to measure distances, weights, volumes, etc. All things which may not come in simple distinct units of measure, which our simple counting numbers would give us. Thus, we need to extend the number concept beyond simply adding another unit and counting higher. For us the key feature in extending the number concept beyond counting, will be through geometry and the concept of the number line. For us, we will think of a number as a point on a number line whose value is the distance our point is away from a fixed point of our choosing on this number line. We call this point the origin and define its location to be the number zero. It is this number line idea, of number being a point on this line whose distance/location from zero is its value, that is required for us to understand all the concepts we will introduce in this book.

If you want to truly grasp the number ideas we present in this course, instead of just memorizing names and facts, then you have to shift away from thinking of numbers as units for counting, and more like distances for measuring. If you can do this, then what we show in this course will make more sense to you. The counting numbers are still there, they are now just points a fixed number of unit distances away from the origin (zero), but now numbers such as fractions, decimals, negative numbers, and irrational numbers will be easier to make sense out of. That being said, we still will use the idea of counting "objects," but now these objects may simply be parts of a whole. This will preserve the idea of number regardless of their type.

The Hindu–Arabic System of Numerals

Now that we have our number concept defined as points on a number line, we next move towards how to easily communicate these values to other people. This requires that we create symbols to represent these numbers. Now this is not as easy as you might think. We have many different types of numbers, but we'd like a single simple system to be able to represent ALL the different types of numbers. Over time the system that was chosen over all other possible ways to do this was something called the Hindu–Arabic Number System. Throughout history there were many different systems developed to represent numbers such as; Egyptian Hieroglyphs, Roman Numerals, Greek Numbers, Mayan Numbers, and Babylonian Numbers, to name a few. However, the one system that won out to be the standard that the entire world adopted, is the Hindu–Arabic Number System.

The Hindu–Arabic Number System is what we call a positional number system. This means that we have symbols we call numerals (or digits) to represent the first nine counting numbers and zero, as well as place–values in our number representation. That means when a numeral (zero through nine) is in what we call our number, where this numeral occurs is important. Its "place" in our number representation, has a value. Since we have ten fingers, it is easy for humans to count in groups of ten, so the Hindu–Arabic Number System counts in groups of ten, and uses place–values using these groups of ten.

Let us begin by reintroducing how we represent these numbers. Our numeral system is called the Hindu–Arabic system, due to its creation in India and preservation and improvement by Arab scholars. It is what is known as a base–ten positional number system. This is because we count in groups of ten, and the location of a numeral within a number is important. It's position is called its place, and that place, we will show, represents a specific value. It is a way to represent all types of numbers very concisely, and has been adopted around the entire world, with only a few minor variations, such as the use of commas and periods within a number representation.

Numerals and Place–Values

First we need our numerals (digits) to represent values of numbers from zero to nine. We are all familiar with the symbols chosen to do so; 0, 1, 2, 3, 4, 5, 6, 7, 8, 9. Next, we define the value of the place (location) our numeral falls in. We have also chosen to write our number representations from left to right, with the leftmost numeral being in the largest place–value and the rightmost numeral being in the smallest (for now) one's location. After the ones, to the left, we have the tens' location. Then we have the hundreds' (or ten groups of ten) location. Next we have the thousands', etc..
For example, to represent the first ten numbers from zero to nine we simply write a single symbol for that number: 0, 1, 2, 3, 4, 5, 6, 7, 8, 9. To write the next largest number ten, however, we use two of our base symbols, a one and a zero (1, 0). By putting the one and zero together as 10 (one ten and zero ones) we have two symbols whose position (place) matters in our representation of the number

ten. The position of the 1 means we have one group of ten and the position of the 0 means we have no ones in our number. This is what we mean by place–value. The place where our number symbol is located has a designated group value.

Thus, the number 26 means that we have two groups of ten and six ones, which is quite different from 62 which has six groups of ten and two ones. Even though the symbols are the same (2,6), the order in which they appear matters.

We can continue this process to represent even larger numbers so the next place –value is defined as ten groups of ten or one hundred, written as 100. That's a one in the hundreds place, a zero in the tens place, and a zero in the ones place. Continuing in this way we have the place–value chart below:

...	Ten Thousands	Thousands	Hundreds	Tens	Ones
...	Ten groups of a thousand	Ten groups of a hundred	Ten groups of ten	Ten groups of one	One group of one
...	1,0000	1,000	100	10	1

This is a topic students frequently have trouble with, but is extremely important. You need to understand the place–value concept, otherwise numbers will not make sense to you.

You should also know that it is possible to represent a number in several distinct ways. The first is called **standard form**, where we write the number as you are used to seeing it. For example the number 247.

Now, we can also write this number using words, or in **word-form** as, two–hundred and forty–seven. Notice how we use the place–value information in the words.

Alternatively, the **place–value** form of the number becomes; two groups of a hundred, four groups of ten, and seven ones, or simply as, 2 hundreds, 4 tens, and 7 ones.

The number can also be represented in **chart-form** both with and without our numeral symbols for numbers as,

Hundreds	Tens	Ones
2	4	7
• •	• • • •	• • • • • • •

Where each disk in the chart represents one group of the named amount in each of the place–values.

Finally, we can write this number in something called **expanded form**. Expanded form simply writes the full place–value with the appropriate values and expands the number out using addition, for example:

$$247 = 200 + 40 + 7$$
or
$$= 2 \times 100 + 4 \times 10 + 7 \times 1$$

We will find this expanded notation to be useful when we eventually do arithmetic with whole numbers.

Being able to understand and use the various representations of a number, means you have a full understanding of the number concept and how it is symbolized in our Hindu–Arabic number system. You should take the time to understand the different representations of whole numbers before proceeding to arithmetic. It will help you better understand the more difficult concepts we will encounter later on.

Consider the following examples:

EXAMPLES:

1. The number 87 has eight groups of ten and seven ones, or 8 tens, and 7 ones.

   We write this number in words as; eighty–seven.

   In chart form we can represent the number as follows:

Tens	Ones
8	7
● ● ● ● ● ● ● ●	● ● ● ● ● ● ●

   In expanded form this is written as:

$$87 = 80 + 7$$
or
$$= 8 \times 10 + 7 \times 1$$

2. The number 32,459 has three groups of ten thousand, two groups of a thousand, four groups of a hundred, five groups of ten, and nine ones, or 3 ten-thousands, 2 thousands, 4 hundreds, 5 tens, and 9 ones.

We write this number in words as; thirty–two–thousand four–hundred and fifty–nine.

In chart form we can represent the number as follows:

Ten Thousands	Thousands	Hundreds	Tens	Ones
3	2	4	5	9
● ● ●	● ●	● ● ● ●	● ● ● ● ●	● ● ● ● ● ● ● ● ●

Again, each disk represents one group of the named amount in each of the place–values.

In expanded form this is written as:

$$32{,}459 = 30{,}000 + 2{,}000 + 400 + 50 + 9$$

or

$$= 3\times10{,}000 + 2\times1{,}000 + 4\times100 + 5\times10 + 9\times1$$

3. The number 109,386 has one group of a hundred thousand, zero groups of ten thousand, nine groups of a thousand, three groups of a hundred, eight groups of ten, and six ones, or 1 hundred–thousands, 9 thousands, 3 hundreds, 8 tens, and 6 ones.

   We write this number in words as; <u>one–hundred and nine–thousand three–hundred and eighty–six</u>.

   In chart form we can represent the number as follows:

Hundred Thousands	Ten Thousands	Thousands	Hundreds	Tens	Ones
1	0	9	3	8	6
●		● ● ● ● ● ● ● ● ●	● ● ●	● ● ● ● ● ● ● ●	● ● ● ● ● ●

   Again, each disk represents one group of the named amount in each of the place–values.

   In expanded form this is written as:

$$109{,}386 = 100{,}000 + 9{,}000 + 300 + 80 + 6$$

or

$$= 1\times100{,}000 + 9\times1{,}000 + 3\times100 + 8\times10 + 6\times1$$

4. The number 7,326,243 has seven groups of a million, three groups of a hundred thousand, two groups of ten thousand, six groups of a thousand, two groups of a hundred, four groups of ten, and three ones, or 7 million, 3 hundred–thousands, 2 ten–thousands, 6 thousands, 2 hundreds, 4 tens, and 3 ones.

   We write this number in words as; <u>seven–million three–hundred and twenty–six–thousand two–hundred and forty–three</u>.

In chart form we can represent the number as follows:

Millions	Hundred Thousands	Ten Thousands	Thousands	Hundreds	Tens	Ones
7	3	2	6	2	4	3
● ● ● ● ● ● ●	● ● ●	● ●	● ● ● ● ● ●	● ●	● ● ● ●	● ● ●

Again, each disk represents one group of the named amount in each of the place–values.

In expanded form this is written as:

$$7{,}326{,}243 = 7{,}000{,}000 + 300{,}000 + 20{,}000 + 6{,}000 + 200 + 40 + 4$$

or

$$= 7\times 1{,}000{,}000 + 3\times 100{,}000 + 20\times 10{,}000 + 6\times 1{,}000 + 2\times 100 + 4\times 10 + 3\times 1$$

As we begin to introduce the arithmetic of numbers, we will find the different forms of writing numbers to be useful in understanding the various arithmetic operations. This is why it is important that you understand the concept of place–value along with not only its standard form, but its word, place–value, chart, and expanded form too.

To further reinforce this concept a physical number representation and calculating device called an abacus was created. It is really the first calculator.

Abacus: a device used for counting and calculating by sliding small balls or beads along rods or in grooves.

Here are some images of an abacus:

Chinese abacus:

Danish abacus:

There any many different types of abacuses, but their purpose is always the same. An abacus is designed to be able to represent any number in our Hindu–Arabic, base–10, number system. Instead of the numerals zero thru nine, we have beads to represent these numerals (actual objects). Furthermore you have separate rods holding the beads and each rod can be used to represent a specific place–value. When counting or adding numbers with this device, whenever you have ten beads on a rod that are placed together, this is equivalent to one bead in the adjacent rod, which represents the next place–value.

Using an abacus, see if you can represent the same concept we introduced above using the place–value chart as well as the expanded form of a number.

EXAMPLE: Represent the number 25,769 using an abacus.

Note that to show the number in the place–value we simply slid the bead on the rod upward. Also note that we are interpreting the different rods from right to left as being different place–values (one's, ten's, hundred's, etc)

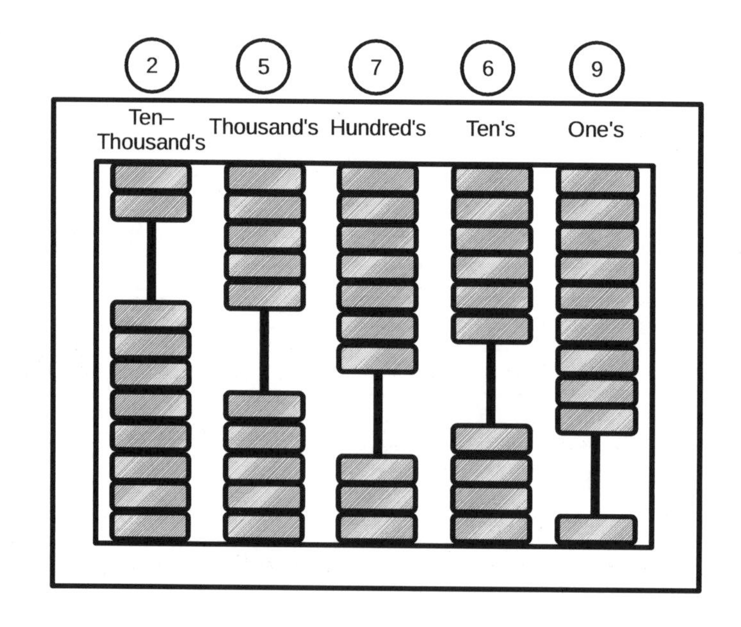

EXERCISES 1.1

Understanding Numbers, Place–Value, and more

Write each number in words, expanded form, using a place–value chart, and on an abacus.

1. 29
2. 36
3. 125
4. 237
5. 3,458
6. 7,239
7. 13,452
8. 79,872
9. 326,913

Given each of the following numbers, identify the place value of the underlined digit.

10. 2,32$\underline{6}$,900
11. $\underline{6}$5,280
12. 1,023,$\underline{6}$77
13. 4$\underline{9}$2,381
14. 82$\underline{1}$,000,352
15. 123,4$\underline{5}$6,789

1.2 ADDING AND SUBTRACTING WHOLE NUMBERS

In the first section of this chapter we introduced the set of natural and whole numbers. Recall, that the natural numbers are our basic counting numbers, and the whole numbers are the counting numbers along with the number zero (many mathematicians include zero with the set of natural numbers and don't define whole numbers differently than naturals, in this presentation we have chosen to distinguish the two sets). Before we introduce the other types of numbers we discussed in the first section, we first want to re–introduce two basic operations we perform on numbers; addition and subtraction.

Adding whole numbers

Eventually we will use a standard algorithm (an algorithm is a predefined step–by–step process) to add two or more numbers together. However, for now we wish to focus on the concept of addition and how it applies to our place–value number system and our number line concept. This way when we use the standard addition algorithm, which tends to mask what we are adding, we will be very aware of what we are actually adding. This will help us to avoid simple mistakes, as we can quickly estimate what the value should be, and if it is not close we can correct our result immediately.

Let's begin with a simple example, 87 + 46. Let's also try not to jump into the standard addition algorithm too soon. There are several ways we can approach this, however, we will just show three. One is more visual, and the other two are more number based.

First let's write these numbers in our place–value charts.

Hundreds	Tens	Ones
	8	7
	••••••••	•••••••

Hundreds	Tens	Ones
	4	6
	••••	••••••

We can add these two numbers by simply adding their respective chart representations as either:

Hundreds	Tens	Ones
	8	7
	4	6

Hundreds	Tens	Ones
	● ● ● ● ● ● ● ●	● ● ● ● ● ● ●
	● ● ● ●	● ● ● ● ● ●

Now in the first column we have 7 ones added to 6 ones. This makes one group of ten and three ones. The one group of ten, however, must now be moved over to the tens column or place–value location. This must be added to the 8 groups of tens and 4 groups of tens already there to give 13 groups of tens. This is divided into 10 groups of ten or 100 and 3 groups of ten. The 10 groups of ten are moved over to the hundreds column. Our final answer is 1 group of 100, 3 groups of ten, and three ones, or 133.
This can be done using either disks to represent the quantity in the place value or number symbols to represent the value in the particular place–value, as shown in the charts below.

Hundreds	Tens	Ones
	8	7
	4	+ 6
	+ 1 ◄	(10) + 3
1 ◄	(10) + 3	3
1	3	3

Hundreds	Tens	Ones
	● ● ● ● ● ● ● ●	● ● ● ● ● ● ●
● ◄	● ● ● ● ● ◄	● ● ● ● ● ●
●	● ● ●	● ● ●
1	3	3

This is the process that leads to the standard addition algorithm. The "carrying" process carries either tens, hundreds, thousands, etc., to the next location in the place–value, which is why the standard algorithm works, as shown below in two equivalent ways.

```
   87              1 1
+  46       or      87
    1             + 46
  1               ----
 ----              133
  133
```

The 1's we are carrying are sometimes 10's, or 100's, etc. They are not 1's! This is why the standard algorithm can be confusing. It breaks everything down to simple single digit addition which makes the problem easy, but depending upon the place–value you are adding you could actually be adding very large numbers, which can make it confusing.

The third approach to adding is to use the expanded notation we developed earlier. Considering the same problem as above, where we add 87 and 46. We rewrite each number in expanded form and add their expanded forms together, as shown below.

$87 = 80 + 7$	Rewrite in expanded notation and add
$\underline{+\,46 = 40 + 6}$	
$= 80 + 40 + 7 + 6$	Rearrange
$= 80 + 40 + 13$	Add the ones together
$= 80 + 40 + 10 + 3$	Rewrite in expanded notation
$= 130 + 3$	Add the tens together
$= 133$	Add and write final answer

The approach is to add like values starting with the ones. If by adding the ones together you get something greater than 9 you write that in expanded form (we got 13 and wrote that as 10 + 3) and then you combine the tens. We got 130 and then added the 3 to get 133.

Consider another example: 1,293 + 3,547

$1{,}293 = 1{,}000 + 200 + 90 + 3$	Rewrite in expanded notation
$\underline{3{,}547 = 3{,}000 + 500 + 40 + 7}$	Rewrite in expanded notation
$= 1{,}000 + 3{,}000 + 200 + 500 + 90 + 40 + 3 + 7$	Rearrange
$= 1{,}000 + 3{,}000 + 200 + 500 + 90 + 40 + 10$	Add the ones together
$= 1{,}000 + 3{,}000 + 200 + 500 + 140$	Add the tens together
$= 1{,}000 + 3{,}000 + 200 + 500 + 100 + 40$	Rewrite in expanded form
$= 1{,}000 + 3{,}000 + 800 + 40$	Add the hundreds & thousands together
$= 4{,}840$	Add and write final answer

We finish this section on addition of whole numbers with one final way to visualize the concept using **tape diagrams**. Tape diagrams allow us to see that addition of whole numbers does not just imply adding individual objects, but can also be thought of as adding two distances. This will become important when we consider adding numbers that are fractions or mixed numbers. For now we just introduce the idea, and we only want you to realize that this is another way to think about the concept of addition.

Consider the example below where we show the result of adding a tape that is 275 units long to a tape that is 138 units long.

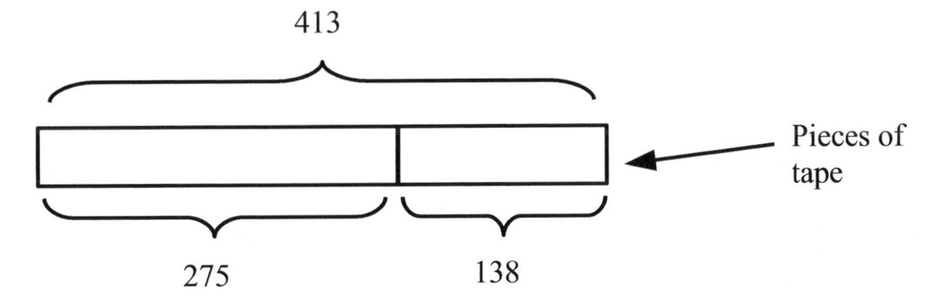

From time to time we will use this aid to better understand the concept we are trying to make sense of. Now we move on to the concept of subtraction.

Subtracting whole numbers

Whereas addition is the process of joining together, subtraction is the process of removing. These two processes are related as being the "opposite" of each other. We will use the same types of representations we did for addition, for illustrating the subtraction process. Only now we remove amounts from a quantity instead of adding them together.

Again, we start with a basic example; 79 – 36

First let's write these numbers in our place–value charts.

Hundreds	Tens	Ones
	7	9
	• • • • • • •	• • • • • • • • •

Hundreds	Tens	Ones
	3	6
	• • •	• • • • • •

We can subtract the second number from the first by simply subtracting or removing the subtracted quantities from respective chart representations as follows:

Hundreds	Tens	Ones
	7	9
	3	6

Hundreds	Tens	Ones
	• • • • • • •	• • • • • • • • •
	• • •	• • • • • •

Now in the first column we have 6 ones subtracted from 9 ones. This leaves us with 3 ones. Then we move on to the tens column and subtract the 3 groups of ten from the 7 groups of ten to get 4 groups of ten. In a similar way we can remove the like values by canceling out the disks in the proper place–value locations.

This is shown below using either disks to represent the quantity in the place value or number symbols to represent the value in the particular place–value:

Hundreds	Tens	Ones
	7	9
	– 3	– 6
	4	3

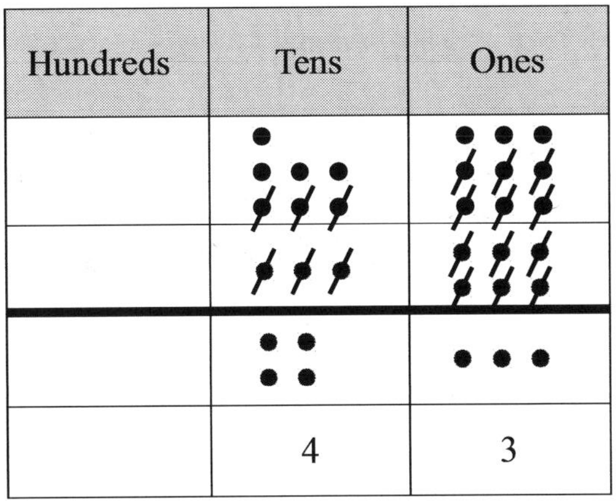

Hundreds	Tens	Ones
	4	3

Consider a second example; 435 – 168. In this case, however, we will need to subtract more than we have in a particular place–value. This leads to the concept of "borrowing." Consider the process using the numerals and then disks illustrated in the figures below:

Hundreds	Tens	Ones
4	**3**	**5**
	2 + 1	
3 + 1	2	10 + 5 = 15
	10 + 2 = 12	**– 8**
3	**– 6**	7
– 1	6	
2		
2	6	7

Since we cannot take 8 away from 5, we must borrow a ten from the tens column to make this 15, which we can take 8 away from to get 7. Now the tens column has 2 tens and we cannot take 6 tens away from 2 tens, so we must borrow a hundred, which is 10 tens, from the hundreds column to make this 12 tens. We can take 6 tens way from 12 tens to get 6 tens. The hundreds column has become 3 hundreds and we can take 1 hundred away to get 2 hundreds, for a final answer of 267.

	Hundreds	Tens	Ones
435			
– 168			
	2	6	7

Using the disks to represent the various groupings, we see the same process at work. We try to subtract the 8 ones from the 5 ones and this cannot be done, so we must borrow a disk from the tens column and this comes over as 10 ones. We then have 15 disks in the ones column and we can subtract (cancel out) the 8 disks from this to get 7 disks left. This has left 2 disks in the tens column, but we are subtracting 6 disks from that. We cannot, so we borrow (move) a disk from the hundreds column to the tens column and it comes over as 10 tens (or ten disks). This now gives us the 2 disks that were already there, plus the 10 disks that were moved over giving a total of 12 disks representing 12 tens. We can subtract (cancel) the 6 disks from the 12 disks in the tens column, leaving 6 disks or 6 tens in the tens column. Finally the disk that was moved over from the hundreds column, now leaves 3 disks and we can subtract the one disk from that to give us 2 disks, or 2 hundreds in the hundreds column. Our final answer is just 267.

This is the process that leads to the standard subtraction algorithm. The "borrowing" process borrows either tens, hundreds, thousands, etc., from the next location to the left for the current place–value you are subtracting in, which is why the standard algorithm works, as is shown below.

$$\begin{array}{r} {}^{3}\;{}^{12}\;{}^{1} \\ \not{4}\;\not{3}\;5 \\ -\;1\;6\;8 \\ \hline 2\;6\;7 \end{array}$$

The 1's we are borrowing are sometimes 10's, or 100's, etc. They are not 1's! This is why the standard algorithm can be confusing. It breaks everything down to simple single digit subtraction, which makes the problem easy, but depending upon the place–value you are subtracting from you could actually be subtracting very large quantities and not just single digits, which can make it confusing. The third approach to subtracting is to use the expanded notation we developed earlier.

Considering the first subtraction problem from above, where we subtracted 36 from 79, or 79 – 36. We rewrite each number in expanded form and subtract their expanded forms from one another, as shown below.

$79 = \quad 70 + 9$	Rewrite in expanded notation
$\underline{-36 = -(30 + 6)} = -30 - 6$	Subtract each place value
$= 70 + 9 - 30 - 6$	Subtract 36 from 79
$= 70 - 30 + 9 - 6$	Rearrange
$= 40 + 3$	Subtract the ones and the tens separately
$= 43$	Add and write final answer

The approach is to subtract like values starting with the ones. However, if you cannot subtract the ones because what you are subtracting is larger, then you will have to alter the approach as shown in the next example.

Consider the example: 435 – 168

$435 = \quad 400 + 30 + 5$	Rewrite in expanded notation
$\underline{-168 = -(100 + 60 + 8)} = -100 - 60 - 8$	Subtract each place value
$= 400 + 30 + 5 - 100 - 60 - 8$	Subtract 168 from 435
$= 400 - 100 + 30 - 60 + 5 - 8$	Rearrange
$= 400 - 100 + 20 + 10 - 60 + 5 - 8$	Borrow 1 ten from 30 for the ones
$= 400 - 100 + 20 - 60 + 10 + 5 - 8$	Rearrange
$= 400 - 100 + 20 - 60 + 15 - 8$	Add the 10 and 5 together
$= 400 - 100 + 20 - 60 + 7$	Subtract 8 from 15
$= 300 + 100 - 100 + 20 - 60 + 7$	Borrow 1 hundred from 4 hundred
$= 300 - 100 + 100 + 20 - 60 + 7$	Rearrange
$= 300 - 100 + 120 - 60 + 7$	Add the 100 and the 20
$= 300 - 100 + 60 + 7$	Subtract 60 from 120
$= 200 + 60 + 7$	Subtract 100 from 300
$= 267$	Add and write final answer

We finish this section on subtraction of whole numbers with one final way to visualize the concept using **tape diagrams**. Tape diagrams allow us to see that subtraction of whole numbers does not just imply subtracting individual objects, but can also be thought of as subtracting (or removing) distances. This will become important when we consider subtracting numbers that are fractions or mixed numbers. For now we just introduce the idea, and we only want you to realize that this is another way to think about the concept of subtraction.

Consider the example below where we show the result of subtracting a tape that is 138 units long from a tape that is 413 units long. This gives us a tape that is 275 units in length as shown below.

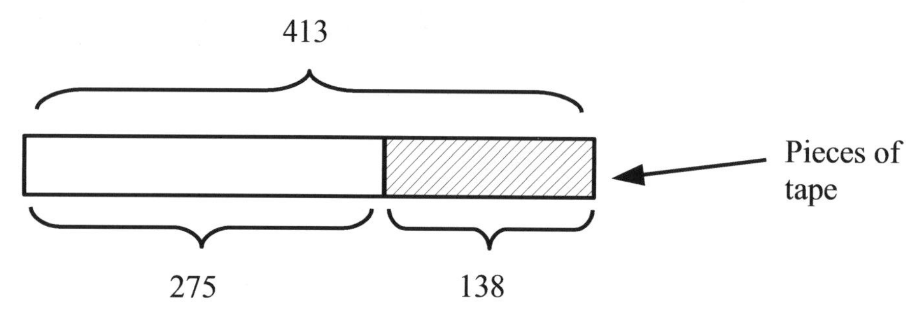

In summary, we have shown a few new conceptual ways to look at both addition and subtraction. In particular, we saw how you can perform both processes using place–value charts with disks or numbers. For addition we showed what carrying really means, and the same was true for illustrating borrowing in subtraction. To further instill these concepts we showed the same processes using the expanded form of a number approach. All this was to justify the standard algorithm for both addition and subtraction. We then ended by showing the visual aspect of what we are doing. The visual aspect is more closely aligned with thinking of our numbers as distances, and we want to reinforce this view.

Try the following exercises, but keep in mind that eventually you will be using the standard algorithm. By doing the exercises involving the charts and expanded form you are only trying to reinforce why the standard algorithm works. They show you conceptually what is happening. The concept behind what you are doing is always more important than the simplified algorithm which often masks what you are really doing.

Applications: The same reasoning can be applied when adding or subtracting other attributes as shown in the following:

EXAMPLE: You have to transport 3 boxes whose weights are given as 16 pounds, 11 pounds, and 32 pounds. What is the total weight of the boxes?

To find the total weight we simply add the weights together, or

$$\text{Total Weight} = 16 \text{ pounds} + 11 \text{ pounds} + 32 \text{ pounds} = 59 \text{ pounds}$$

EXAMPLE: There were 148 people on the plane. 37 people got off at the last airport, but 11 additional people came on. How many people are now on the plane?

$$\text{Total} = 148 \text{ people} - 37 \text{ people} + 11 \text{ people} = 122 \text{ people}$$

EXERCISES 1.2

Adding and Subtracting Whole Numbers

In each of the following problems, add or subtract using the place value chart and the standard algorithm.

1. $85+52$
2. $343+977$
3. $168+452$
4. $743+267$
5. $85-52$
6. $977-343$
7. $452-168$
8. $743-267$

In each of the following problems, add or subtract using the expanded form and the standard algorithm.

9. $69+56$
10. $472+481$
11. $279+642$
12. $2{,}435+1{,}854$
13. $69-56$
14. $472-281$
15. $642-279$
16. $2435-1854$

In each of the following problems, add or subtract using the tape diagram .

17. $85+52$
18. $343+977$
19. $168+452$
20. $743+267$
21. $85-52$
22. $977-343$
23. $452-168$
24. $743-267$

Applications

25. Mary purchased 240 markers. Peter purchased 93 markers. How many markers did the two of them purchase together?

26. Kevin has 70 marbles. Tommy has 18 less marbles that Kevin. How many marbles does Tommy have?

27. There were 2,500 people who attended a football game. After half time, 843 people left the game, how many people are left in the stadium?

28. The bookstore has in stock 548 English textbooks, 402 Science textbooks, 610 Mathematics textbooks, and 389 Social Science textbooks. How many textbooks does the bookstore have total?

29. Madeline baked 48 muffins on Monday, 27 muffins on Tuesday, and 39 muffins on Wednesday. How many muffins did she bake during the three days?

30. Ana has 230 stickers. Her brother Jonathan has 178 stickers. How many more sticker does Ana have than Jonathan?

1.3 MULTIPLYING WHOLE NUMBERS

In the previous section we reintroduced addition and subtraction focusing more on the concepts behind the standard algorithms. In this section we take the same approach with multiplication and division. The focus here is on a conceptual understanding of what is behind the standard algorithms and not necessarily practice with the standard algorithms themselves. This is material you have seen already. Now, however, we want you to approach it as if you have not, and try and grasp the concepts behind the algorithms you may have used blindly for years. If you understand the concept, then the standard algorithm makes more sense and leads to fewer mistakes and misunderstandings when you use it.

The key concept behind multiplication is that it is simply a repeated addition. To better understand this concept we show the process using something called a tape diagram, then to get a better understanding of our place–value system, we show the place–value approach. Next we show the expanded form approach, and the partial products (or box) approach, and then we end with the standard algorithm. This may seem like a lot of different ways to multiply, and it is, but the purpose is to focus on the concept of multiplication early on, and see how the standard multiplication algorithm really works. In the end you will be using the standard algorithm for multiplication, but if you understand where the algorithm comes from and why and how it works, you can use it more effectively.

Let's illustrate the first two approaches with a simple example. The other approaches are not applicable when we are just multiplying single digit numbers, so we'll illustrate them with a more complicated example later.

Consider 2×6 . This means we have 2 groups of 6 or we are adding 6 to itself 2 times. (Note: We could also view this as adding 2 to itself 6 times, but for now we'll follow the convention that the first numbers is how many times we repeat the second number. Typically we like to have the smaller number first to demonstrate the concept.)

The first approach helps us to visualize what we are trying to find. We do this with a **tape diagram**, which is similar to our number line.

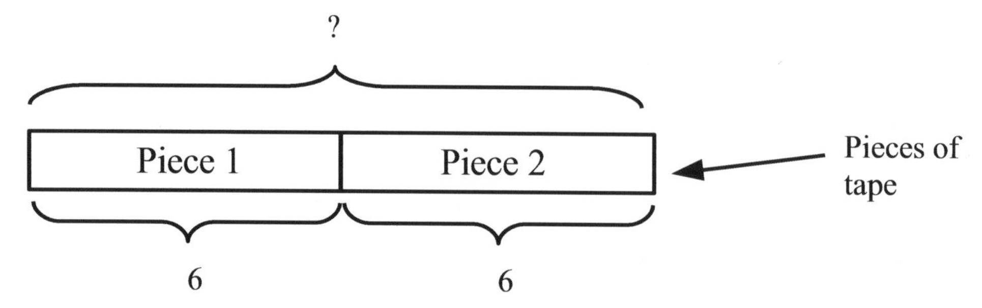

We are trying to find the result of adding 6 to itself 2 times, or $2\times6 \;=\; 6+6=12$.

The next approach is to write the repeated additions in our **place–value chart**, and perform the additions as we did in the last section.

	Tens	Ones
6		• • • • • •
6		• • • • • •

Doing the addition we learned in the last section we have:

	Tens	Ones
6	•	• • • • • •
6		• • • • • •
12	•	• •

So $2 \times 6 = 12$

Now we consider a more complicated example and show the other approaches.
Consider the example: 3×23

Tape Diagram
The diagram shows us the 3 repeated additions of 23 and what we are trying to find.

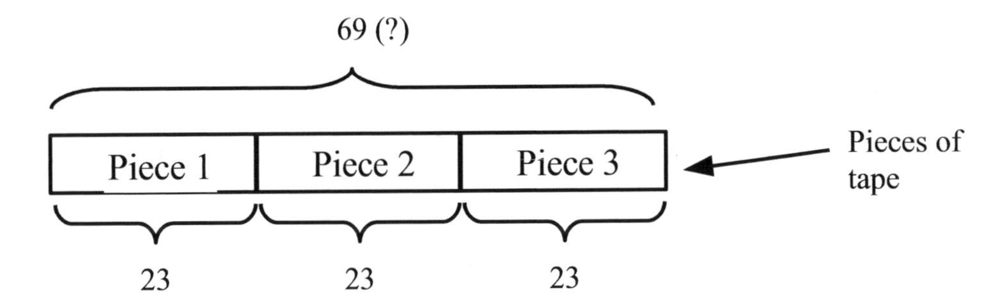

Place–value chart
Using our place–value chart we have:

	Tens	Ones
23	● ●	● ● ●
23	● ●	● ● ●
23	● ●	● ● ●

Doing the additions:

	Tens	Ones
23	● ●	● ● ●
23	● ●	● ● ●
23	● ●	● ● ●
69	● ● ● ● ● ●	● ● ● ● ● ● ● ● ●

There are 6 tens and 9 ones so the results is 69, or $3 \times 23 = 69$

We now introduce three new techniques.

<u>Expanded Form</u>
Another way to look at this is using the **expanded form** of the number. In this example we use the expanded form with repeated additions, and then also introduce the distributive property approach. (Note: The distributive property has not been discussed previously, but it is actually one of the axioms of our system of numbers.)

We start by rewriting 23 in expanded form: $23 = 20 + 3$

First, we shall add 23 to itself 3 times in its expanded form.

$3\times23 = (20 + 3) + (20 + 3) + (20 + 3)$	Rewrite in expanded form
$= 20 + 20 + 20 + 3 + 3 + 3$	Rearrange
$= 60 + 9$	Add like place-values
$= 69$	Add the final numbers

Alternatively we can also use a similar approach, but this time we use what you've learned in previous courses called the distributive property.

Multiplying by 3 and using the distributive property we have:

$3\times23 = 3\times(20 + 3) = 3\times20 + 3\times3$	Distribute the 3 to both the 20 and the 3
$= 60 + 9$	Do the multiplications
$= 69$	Add the final numbers

Partial products

The next approach is really just another variation of the expanded from approach with the distributive property, but tends to be a little more visual. This is called the method of **partial products**.

We begin by setting up a chart (or boxes) with the numbers we are multiplying broken down into standard form. With one of the numbers on top, and the other along the left side as show in the figure below.

	20	3
3	60	9

We then multiply the side term by the top term (usually very easy multiplications) and place the result in the connecting box. When completed, we add the connecting boxes to get our answer. In this case it is the 60 added to the 9 to get 69.

Standard Algorithm

We now introduce the **standard algorithm**. The standard algorithm is what we will use in the end, but sometimes the problems can be quite complicated and we may lose sight of what we are multiplying. It is when this happens that we can fall back on the above techniques to conceptually correct any mistakes we might make using this algorithm.

In the standard algorithm we multiply the 3 ones times the 3 ones to get 9 ones, and place the result in the ones column. Next we multiply the 3 ones times the 2 tens to get 6 tens and place the result in the tens column. Finally we add the 60 and the 9 together to get 69. This is illustrated below.

$$\begin{array}{rr} & 2\ 3 \\ \times & 3 \\ \hline & 9 \\ & 6\ 0 \\ \hline & 6\ 9 \end{array} \qquad \begin{array}{l} \\ \\ 3\times3 \text{ ones} \\ 3\times2 \text{ tens} \\ \\ \end{array}$$

Typically we don't write the extra multiplications and streamline it to just.

$$\begin{array}{r} 23 \\ \times \quad 3 \\ \hline 69 \end{array}$$

We have shown how to multiply using the tape diagram, the place–value chart, partial products, the expanded form, and the standard algorithm. Let's consider other examples using all these techniques:

EXAMPLE 1: Multiply 4×56

Tape diagram
Visually with the tape diagram this becomes.

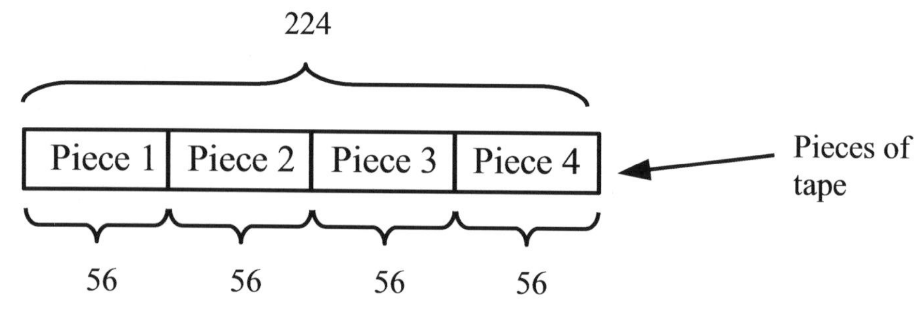

Showing us that 4 copies of 56 equals 224.

Place–value chart
4 repeated additions of 56

Hundreds	Tens	Ones
	● ● ● ● ●	● ● ● ● ● ●
	● ● ● ● ●	● ● ● ● ● ●
	● ● ● ● ●	● ● ● ● ● ●
	● ● ● ● ●	● ● ● ● ● ●

Add the ones and move groups of 10 ones over to the tens column

Hundreds	Tens	Ones
	• • • • • •	• • • • • •
	• • • • • •	• • • • • •
	• • • • •	• • • • • •
	• • • • •	• • • • • •

Add the tens and move groups of 10 tens over to the hundreds column. Sum the final result.

Hundreds	Tens	Ones
•	• • • • • •	
•	• • • • • •	
	• • • • •	
	• • • • •	• • • •
• •	• •	• • • •
2	2	4

Expanded notation

Now, using the expanded form, we write 56 as 50 + 6 and use repeated additions to find,

$$\begin{aligned}4\times56 &= (50+6)+(50+6)+(50+6)+(50+6)\\ &= 50+50+50+50+6+6+6+6\\ &= 200+24\\ &= 224\end{aligned}$$

or using the distributive property to obtain the same answer

$$\begin{aligned}4\times56 = 4\times(50+6) &= 4\times50+4\times6\\ &= 200+24\\ &= 224\end{aligned}$$

Partial products

	50	6
4	200	24

$$4 \times 56 = 200 + 24 = 224$$

Standard algorithm

Finally, to solve this problem using the standard algorithm we have:

$$\begin{array}{rrr} & 5 & 6 \\ \times & & 4 \\ \hline & 2 & 4 \\ 2 & 0 & 0 \\ \hline 2 & 2 & 4 \end{array} \quad \begin{array}{l} \\ \\ 4 \times 6 \text{ ones} \\ 4 \times 5 \text{ tens} \\ \\ \end{array}$$

Note that you could have multiplied 4 × 5 tens first then 4 × 6 ones, and then add the partial products. As long as your alignment is correct, the answer would be the same.

Here is the streamlined approach:

$$\begin{array}{r} {}^{2} \\ 56 \\ \times \quad 4 \\ \hline 4 \\ +\,22 \\ \hline 224 \end{array}$$

What happens when you have to multiply a two-digit number by a two–digit number?

Multiplying two–digit by two–digit numbers.

EXAMPLE 2: Multiply 47×26.

We could show the result using the tape diagram or the place value chart, but having to write 26 times 47 of pieces of tape or disks on the chart is not practical. So we shall just use the expanded form, partial products, and the standard algorithm.

Expanded form

Since both numbers are fairly large, we will use the distributive property approach instead of repeated additions.

$47 \times 26 = 47 \times (20 + 6)$	Write 26 in expanded form
$= 47 \times 20 + 47 \times 6$	Distribute the 47
$= (40 + 7) \times 20 + (40 + 7) \times 6$	Now, we can write 47 in expanded form
$= 40 \times 20 + 7 \times 20 + 40 \times 6 + 7 \times 6$	We can distribute backwards
$= 800 + 140 + 240 + 42$	Perform multiplication
$= 1{,}222$	Adding all partial products we get

Partial products

In expanded form 47 is 40 + 7 while 26 is 20 + 6. We created a grid and set it up as below. Then multiply each number in the horizontal boxes by each number in the vertical boxes to get 4 partial products. Add these 4 partial products and that is the product of the two given numbers.

	20	6
7	140	42
40	800	240

$$47 \times 26 = 800 + 140 + 240 + 42$$
$$= 1{,}222$$

Standard algorithm

Using the standard algorithm, we get the partial products as shown below, and add them in their appropriate columns. You must pay careful attention to the place–value of each digit when doing this.

4 7		
× 2 6		
4 2	6 × 7 ones,	6 × 7 = 42
2 4 0	6 × 4 tens,	6 × 40 = 240
1 4 0	2 tens × 7 ones,	20 × 7 = 140
8 0 0	2 tens × 4 tens,	20 × 40 = 800
1, 2 2 2		

In streamlined form:

```
    1
    4
    47
 × 26
-----
     2
   28
    4
 + 9
-----
1,222
```

You can begin to see how the standard algorithm can get confusing if you lose track of what you are multiplying and adding!

EXAMPLE 3: Multiply 345×29. In this last example we will only use the standard algorithm. In this example, we show that you can start with the leftmost digit of a number, perform the multiplication and still end up with the correct answer as long as the proper place value is considered.

```
        3 4 5
 ×        2 9
--------------
      6 0 0 0    2 tens × 3 hundreds, 20 × 300 = 6000
        8 0 0    2 tens × 4 tens, 20 × 40 = 800
        1 0 0    2 tens × 5 ones, 20 × 5 = 100
      2 7 0 0    9 ones × 3 hundreds, 9 × 300 = 2700
        3 6 0    9 ones × 4 tens, 9 × 40 = 360
          4 5    9 ones × 5 ones, 9 × 5 = 45
      2 1
------------------------------------------------------
   1 0, 0 0 5
```

345×29 = 10,005

In streamlined form:

```
       1
      4 4
      3 4 5
    × 2 9
    ---------
          5
        0
    3 1
        0
      9
  + 6
  -----------
  1 0, 0 0 5
```

We end with one last example showing how partial products can be done using either 2 or 4 partial products.

EXAMPLE 4: Solve 26×35 using four partial products and two partial products.

Four partial products

In expanded form 26 is 20 + 6 while 35 is 30 + 5. We created a grid and set it up as below. Then multiply each number in the horizontal boxes by each number in the vertical boxes to get 4 partial products. Add these 4 partial products and that is the product of the two given numbers.

	20	6
5	100	30
30	600	180

$$
\begin{aligned}
26 \times 35 &= 100 + 600 + 30 + 180 \\
&= 910
\end{aligned}
$$

Two partial products

We can merge the bottom two boxes on the left (or top two if we want) and perform the partial products. We get the same result doing it this way as well.

$$
\begin{aligned}
20 \times 35 &= 20 \times (30 + 5) \\
&= 20 \times 30 + 20 \times 5 \\
&= 600 + 100 \\
&= 700
\end{aligned}
$$

$$
\begin{aligned}
6 \times 35 &= 6 \times (30 + 5) \\
&= 6 \times 30 + 6 \times 5 \\
&= 180 + 30 \\
&= 210
\end{aligned}
$$

	20	6
35	700	210

$$26 \times 35 = 700 + 210$$
$$= 910$$

EXERCISES 1.3

Multiplying Whole Numbers

Solve the following multiplication problems with a tape diagram, place–value chart with disks, expanded notation, partial products, and the standard algorithm.

1. 3×24
2. 3×42
3. 4×34
4. 4×27
5. 5×42

6. Cindy says she found a shortcut for doing multiplication problems. When she multiplies 3×24, she says, "3×4 is 12 ones, or 1 ten and 2 ones. Then, there's just 2 tens left in 24, so add it up, and you get 3 tens and 2 ones." Do you think Cindy's shortcut works? Explain your thinking in words and justify your response using a model or partial products.

7. Solve 26×34 using 4 partial products and 2 partial products. Remember to think in terms of units as you solve. Write an expression to find the area of each smaller rectangle in the area model.

8. Solve 26×34 using 4 partial products and 2 partial products. Remember to think in terms of units as you solve. Write an expression to find the area of each smaller rectangle in the area model.

9. Solve 52×26 using 2 partial products. Match each partial product to its area on a model.

Solve the following using the standard algorithm. Visualize using the area model to help you.

10. 52×26
11. 37×24
12. 152×73
13. 352×385

Applications

14. There are 15 cookies in each box of cookies. Pam has 8 boxes of cookies. How many cookies does Pam have?

15. There are 7 bags of candy. Each bag has 48 pieces of candies. How many pieces of candies are there in the 7 bags?

16. There are 30 students in each classroom on one floor. There are 17 classrooms on the floor. How many students are there on the floor at that time?

17. There are 4 dry erase markers in each box. Nathan bought 17 boxes of dry erase markers. How many dry erase markers did Nathan buy? If each box costs $5 to buy, how much did Nathan pay for the 17 boxes of dry erase markers.

1.4 DIVIDING WHOLE NUMBERS

Division is the opposite of multiplication. In multiplication we are looking to find the repeated additions. In division we still have the concept of repeated additions, but we are now looking for the number we are adding, and not the sum of the repeated additions.

We can see this by looking at the tape diagram as we did for multiplication. Now consider the simple example $12 \div 2$

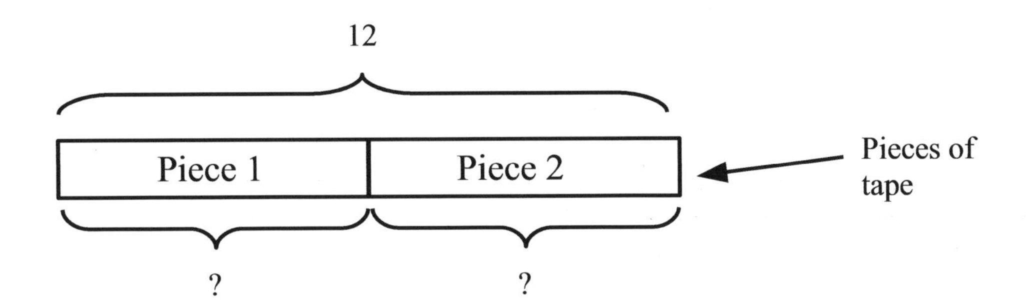

Now, instead of finding the value 12 that we did for multiplication (the sum of the repeated additions), we are now trying to find out the size of the group we would get if we broke 12 into 2 groups (what we are adding).

Just as we had several approaches to doing multiplication, we have several for division as well. These approaches are mainly to help us understand the concept better so that when we use the standard division algorithm called long division, we will know what we are actually doing, and not just mindlessly following steps.

The approaches we will introduce for division are the tape diagram, the place–value chart, an area model, and the long division algorithm.

We shall demonstrate these approaches through some examples.

EXAMPLE 1: Divide 6 ones by 3, that is $6 \div 3$

<u>Tape diagram</u>

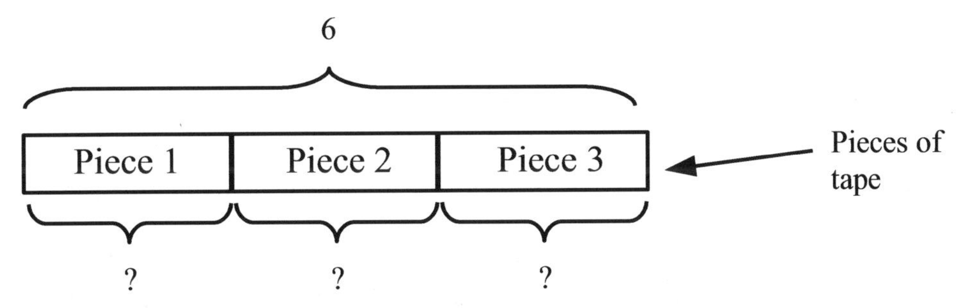

From the picture above we can see the answer is 2.

Place–value chart

We display or place 6 disks on the place value chart. The divisor is 3 so we draw 3 groups below the 6 disks. Then we distribute one disk to each group. Cross off the ones that have been distributed to each group. Continue this process until we observe how many disks are in each group. In this example, it is 2 which means 6 divided by 3 gives us 2. There is no remainder for this problem.

Tens	Ones
	●●●●●●

Tens	Ones	
	●●●●●●	
	●●	2 ones
	●●	
	●●	

EXAMPLE 2: Divide 36 ones by 3, that is 36 ÷ 3

Tape diagram

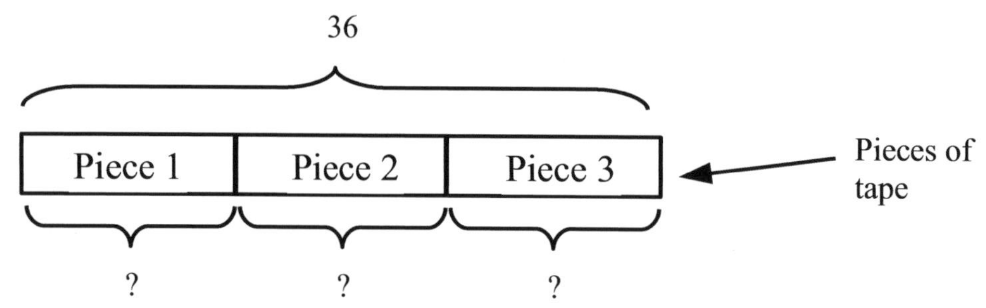

From this picture we can see that the group size must be 12.

Place–value chart

We create the place value chart with 3 tens and 6 ones to represent the number 36. Then draw 3 groups below and distribute each of the 3 tens and 6 ones making sure that there is enough for each group to have the same number of disks. If you are able to achieve this, that means there is no remainder, otherwise, there is a remainder. We will work on an example with a remainder later.
This is what the place–value looks like for the number 36,

	Tens	Ones
36	●●●	●●●●●●

We can evenly divide the number 36 into three equal groups of 12 as shown below:

	Tens	Ones	
36	●●●	●●●●●●	
	●	●●	3 groups of 1 ten and 2 ones so 36 ÷ 3 = 12.
	●	●●	
	●	●●	

<u>Area model</u>
We now show an alternative approach to a division problem by introducing something called an area model. This comes from that fact that when we multiply two distances together we get an area. Thus, 3 meters by 6 meters gives us a rectangular area of 18 meters^2. In multiplication we are asked to find the 18 meters^2, but in division we are asked to find one of the lengths of this rectangular area, given it area and one of the lengths of the rectangle.

This is a different approach of doing division that is more visual. We'll show a couple of examples to show how it is done..

EXAMPLE 3: Divide: 10 ÷ 2.

Let's use grid paper to draw a rectangle with an area of 10 square units and one side length of 2 units. To find the unknown side length, the area is 10, so we know the unknown side is 5 units. If the width is 2 units, that means the length is 5 units, and 2 units times 5 units gives an area of 10 square units. We can count and mark off by twos until we get to 10.

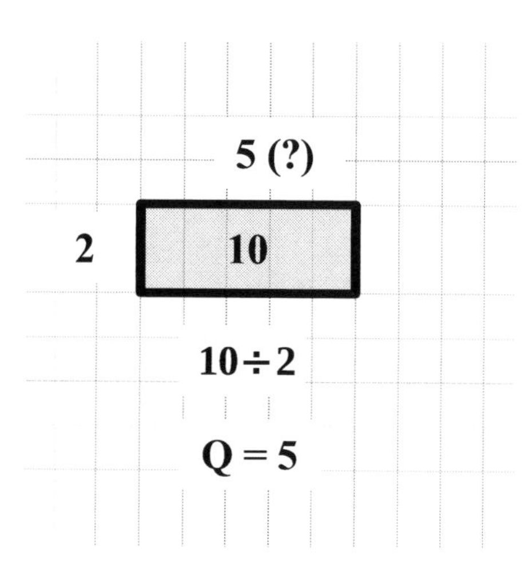

Q means quotient which is the result of division. So **10÷2 = 5**

Note: Units in this example could be centimeters, inches, or other units. We just use units which is generic term.

Oftentimes when we do division the number of groups we are dividing our number into may not break our number (area) into equal portions. There may be some left over amount that we have to account for. We show how to do this with our area model below.

EXAMPLE 4: Divide: 11 ÷ 2.

We do as we did with the example above. But notice that now we have 11, instead 10, that is divided by 2. Eleven square units is the total area. Let's draw a rectangle starting with a width of 2 units. We'll continue lengthening it until we get as close to 11 square units as we can.

A length of 5 units and width of 2 units is as close as we can get to 11 square units.

We can't do 2 × 6 because that's 12 square units and the total area is 11 square units. We can show a total area of 11 square units by modeling 1 more square unit. The remainder of 1 represents 1 more square unit.

Another way is to start with groups of two boxes and keep adding them until if we added two more we would be greater than 11. Then count the remainder to get to 11. In this case we would have 5 groups of 2 with a remainder of 1.

This is equivalent to writing the mathematical sentence as:

$$(2 \times 5) + 1 = 11$$

See the figure below to illustrate the procedure.

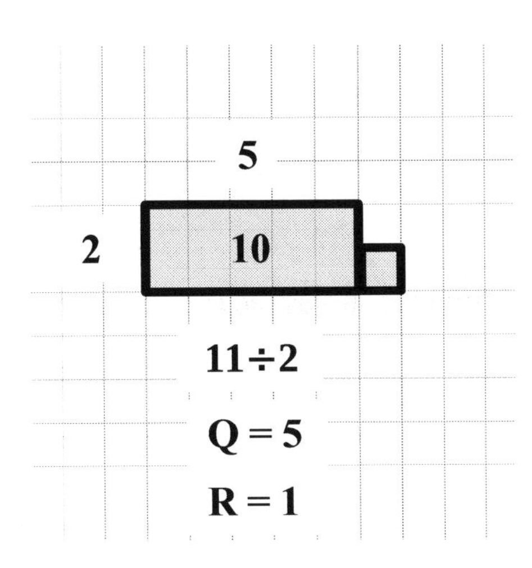

EXAMPLE 5: Divide: $38 \div 4$.

We want to divide 38 by 4. We start by grouping 4 and keep going until we see that once we get to 9 groups of 4, we have 36. And 2 left which is not enough to group another 4.

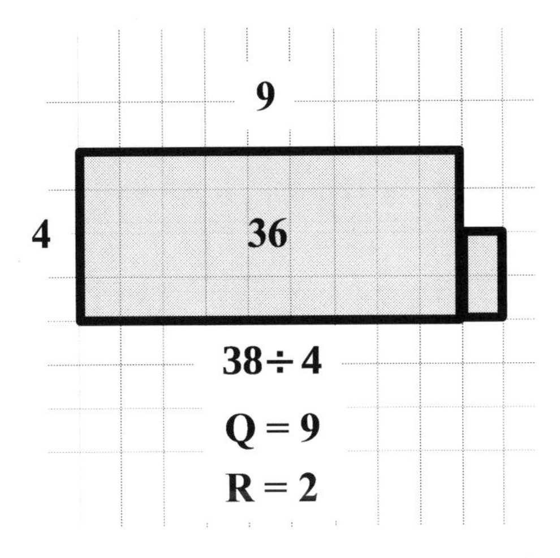

So the area diagram shows that $38 \div 4 = 9$ with a remainder of 2. We can write this in a sentence format as:

$$(4\times 9) + 2 = 38.$$

<u>Standard algorithm</u>
We now introduce the standard algorithm for division called **long division**.

Let's illustrate this using our example $38 \div 4$.

We first write this as division problem using the long division notation:

$$4\overline{)\,38}$$

The process proceeds as follows. We first look for the largest number of 4's that go into 38 and that is 9, since $4 \times 9 = 36$ and $4 \times 10 = 40$ and 40 is too large. So we place the 9 as shown below.

$$4 \overline{)\,38}^{\;9}$$

We then do the multiplication $4 \times 9 = 36$ and subtract that value from 38 as follows:

$$\begin{array}{r} 9 \\ 4 \overline{)\,38} \\ -36 \\ \hline 2 \end{array}$$

Now since 2 cannot be divided into 4 groups evenly (as a whole number amount) we stop and write our answer as 9 remainder 2.

More Division of Whole Numbers – with larger single digit divisors

We have previously learned division by small single digits numbers. We now look at dividing by larger numbers.

EXAMPLE 1: We all know there are 7 days in a week. How many weeks are in 259 days?

This means we want to know how many 7 days do we have in 259.

Tape diagram

We can visualize what we are trying to obtain with a tape diagram

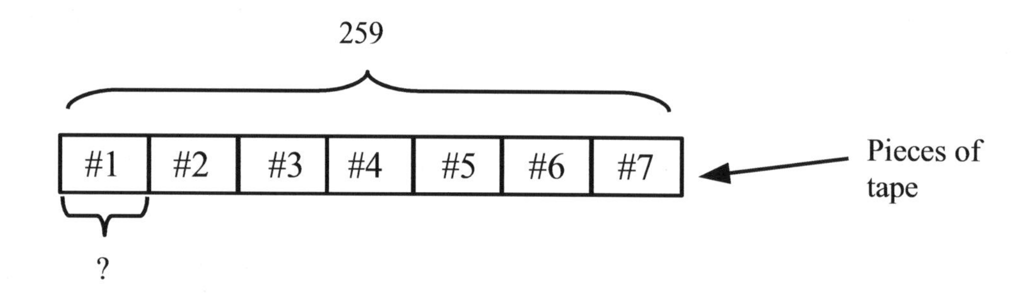

We are breaking 259 into 7 equal sized groups and we want to determine the size of these groups. Here the tape diagram is used more as a visual image of what we are trying to find and not as an easy way of finding it. For that we look at the other approaches.

Place–value chart

Representing the number 259 in our place–value chart, it looks like the following.

Hundreds	Tens	Ones
●●	●●●●●	●●●●● ●●●●

We now need to divide the 259 into 7 even sized groups and determine how many is in each group.

Hundreds	Tens	Ones
(●●) →	●(●●●●) → ●●●●●●●●●● ●●●●●●●●●●	●●●●● ●●●● ●●●●●●●●●● ●●●●●●●●●● ●●●●●●●●●● ●●●●●●●●●●
	●●●	●●●●●●●
	●●●	●●●●●●●
	●●●	●●●●●●●
	●●●	●●●●●●●
	●●●	●●●●●●●
	●●●	●●●●●●●
	●●●	●●●●●●●

Notice that 2 hundreds does not have enough disks to distribute to the seven groups. So we can turn the two hundreds into twenty tens and place the disks in the tens column.

There are 25 disks in the tens digit. If we distribute the disks uniformly into the 7 groups, we will have 3 in each group. This leaves us with 4 tens which is not enough to be distributed in the 7 groups.

So we make the 4 tens into 40 ones and place the disks in the ones column. Now we have a total of 49 ones which can be distributed evenly in the 7 groups. Each group will get 7 disks each.

There is no remainder in this problem so we conclude that $259 \div 7 = 37$. To check our work, we multiply 7×37, and we get 259. There are 37 weeks in 259 days.

Now, as the divisor (the number of groups we are dividing our number into) gets larger, it is harder to use the place–value chart or even the area model to determine the answer. Thus, the long division algorithm can be used to obtain the same answer using a simpler, less involved, but equivalent algorithm.

<u>Long division</u>
First we write the problem using the long division notation.

$$7\overline{)\,259}$$

Next we look at the 7 and the 2 and see how many times does 7 go into 2 evenly. It does not, so we then look at the next digit or 25. 7 goes into 25 at most 3 times so we write the 3, and then multiply the 7 by 3 to get 21 and subtract the 21 from 25 to get 4 as shown below:

$$\begin{array}{r} 3 \\ 7\overline{)\,259} \\ -21 \\ \hline 4 \end{array}$$

We now bring down the 9

$$\begin{array}{r} 3 \\ 7\overline{)\,259} \\ -21\downarrow \\ \hline 49 \end{array}$$

We look to see how many times 7 goes into 49 and it goes in 7 times evenly (no remainder), so we complete the problem as follows:

$$\begin{array}{r} 37 \\ 7\overline{)\,259} \\ -21 \\ \hline 49 \\ -49 \\ \hline 0 \end{array}$$

The answer is 37 with 0 remainder or simply 37.

Hopefully you can see why the algorithm is useful as a simple process to find the answer, but why it is also confusing conceptually as to what you are doing. If you don't put the numbers in the correct place, or if you don't start correctly, it is easy to get lost, and consequently obtain the wrong answer. While the other approaches are harder to write out and implement, they at least mimic the actual division process, where this algorithm does not.

To help get the correct answer using this algorithm you need to remember that you are not looking for how many times 7 goes into 25 evenly, but really how many times 7 goes into 250 evenly. That answer is 30, which is why we place the 3 where we do. Then when we distribute these 210 object in the 7 groups (30 each) we must subtract them from the total of 259 and we have 49 left over. Now when we distribute these remaining 49 into the 7 groups, we put 7 into each without any left over, so the answer is 30 + 7 or 37.

The long division algorithm is a nice little algorithm, but you have to be extremely careful when using it, as the concept of division is not so obvious when performing the steps!

EXAMPLE 2: Everyone is given the same number of colored pencils in art class. If there are 249 colored pencils and 8 students, how many pencils does each student receive?

This is a division problem; $249 \div 8 = ?$

We can use the place–value chart, but the process now becomes a little more cumbersome, so we opt to use the long division algorithm instead.

First write the long division notation,

$$8\overline{)\,249}$$

Next, find out how many times 8 goes into 24 (really 240), and write the 3, multiply and subtract. In this case we get zero.

$$\begin{array}{r} 3 \\ 8\overline{)\,249} \\ -24 \\ \hline 0 \end{array}$$

Bring down the 9,

$$\begin{array}{r} 3 \\ 8\overline{)\,249} \\ -24\downarrow \\ \hline 9 \end{array}$$

Repeat the process

$$\begin{array}{r} 31 \\ 8\overline{)\,249} \\ -24 \\ \hline 9 \\ -\ 8 \\ \hline 1 \end{array}$$

This problem has a remainder, which we can write as follows:

$$\begin{array}{r} 31\ \ \text{R1} \\ 8\overline{)\,249}\phantom{\text{R1}} \\ -24\phantom{\text{R1}} \\ \hline 9\phantom{\text{R1}} \\ -\ 8\phantom{\text{R1}} \\ \hline 1\phantom{\text{R1}} \end{array}$$

So, we write our final answer as either 31 remainder 1, or 31 R1.

The tape diagram can be used here, just as a visual image of what we found.

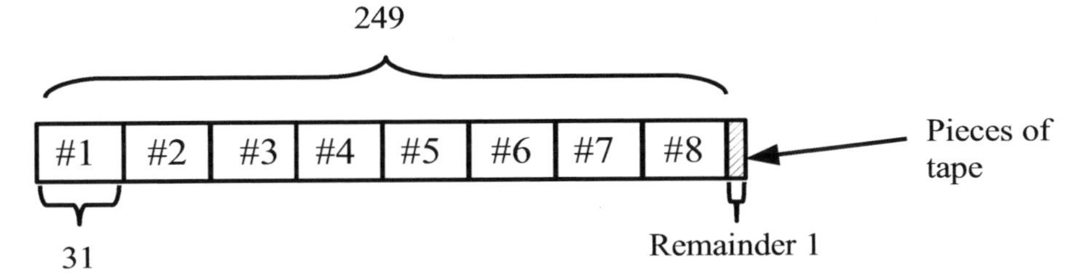

Each student will receive 31 colored pencils. There will be 1 pencil left over. To account for the remaining one pencil, we shade a small portion at the end of the tape diagram to represent the remaining pencil.

So, $8 \times 31 + 1 = 249$ or $249 \div 8 = 31$ with a remainder of 1

 H.W. 1–13 odds

EXERCISES 1.4

Dividing Whole Numbers

Solve the following division problems with the indicated approach.

	Show division using an area model.
1. 24 ÷ 4 Quotient = ________ Remainder = ______	Can you show 24 ÷ 4 with one rectangle? ______
2. 25 ÷ 4 Quotient = ________ Remainder = ______	Can you show 25 ÷ 4 with one rectangle? ______ Explain how you showed the remainder:

Solve using an area model. The first one is done for you.
Example: $25 \div 3$

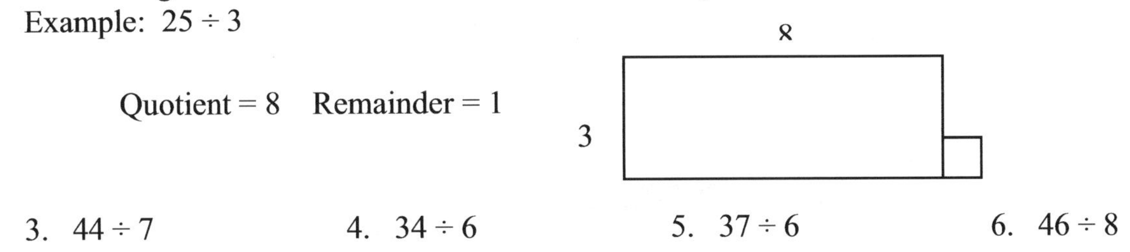

3. $44 \div 7$ 4. $34 \div 6$ 5. $37 \div 6$ 6. $46 \div 8$

Solve the following division problems. Use the long division algorithm and/or draw tape diagrams to help illustrate what you solved. If there is a remainder, shade in a small portion of the tape diagram to represent that portion of the whole.

7. Meneca bought a package of 435 party favors to give to the guests at her birthday party. She calculated that she could give 9 party favors to each guest. How many guests is she expecting?

8. 4,000 pencils were donated to an elementary school. If 8 classrooms shared the pencils equally, how many pencils did each class receive?

9. 2,008 kilograms of potatoes were packed into sacks weighing 8 kilograms each. How many sacks were packed?

10. A baker made 7 batches of muffins. There was a total of 252 muffins. If there was the same number of muffins in each batch, how many muffins were in a batch?

11. Samantha ran 3,003 meters in 7 days. If she ran the same distance each day, how far did Samantha run in 3 days?

12. A boy has collected 260 marbles in a period of time. He wants to give 13 marbles to each of his classmates. How many classmates does the boy have?

13. A farm stand harvested 210 bags of apples. Each bag has 15 apples. How many apples did the owner of the farm stand harvest?

1.5 POWERS OF TEN AND POSITIVE EXPONENTS

Mathematics is a powerful language that often defines useful devices for writing things in a condensed and more easily understood way. One particular example relates to our place-value number system and the powers of ten. In this chapter we start with this place-value example and then show how to extend the notation even further.

When writing the place-values of our number system, we are often required to write larger and larger numbers. For very large place-values the numbers can become quite large and take up quite a bit of space. There is, however, a way to condense how we write these large numbers by introducing a new device called an exponent.

Consider the place value chart below.

...	One Hundred Thousands	Ten Thousands	Thousands	Hundreds	Tens	Ones	Decimal Point
...	100,000	10,000	1,000	100	10	1	.

As we move further to the left in the place-value chart we see that the place-value gets larger and larger, and it becomes increasingly more difficult to represent these larger place-values. It is also difficult to easily distinguish between these place-values. For example consider the two large place-value numbers:

1000000000000000 and 100000000000000

It is hard to see at first glance which number is larger and by how much. Now, we often use commas to help us distinguish the differences, and each number above could be written as:

1,000,000,000,000,000 and 100,000,000,000,000

It is now easier to see that the first number has an extra zero, so it is ten times larger than the second. However, it still takes a large amount of effort to write down each place-value. Our Hindu-Arabic number system has place values based upon factors of ten, which is why it is called a base-10 number system. In other words if we move one place to the left in our system, it is equivalent to multiplying by a factor of ten, as shown from right to left below:

$10 \times 10 \times 10 = 1{,}000$, $\quad 10 \times 10 = 100$, $\quad 10 \times 1 = 10$, $\quad 1$

To capture this behavior in a clear and concise manner, the concept of an exponent was created. An exponent is a super-scripted number that is placed to the right of a number, called the base, and it indicates how many factors of the base the number has in it. Thus,

$10^1 = 10$ One factor of the base 10 number
$10^2 = 10 \times 10 = 100$ Two factors of the base 10 number
$10^3 = 10 \times 10 \times 10 = 1{,}000$ Three factors of the base 10 number
$10^4 = 10 \times 10 \times 10 \times 10 = 10{,}000$ Four factors of the base 10 number
$10^5 = 10 \times 10 \times 10 \times 10 \times 10 = 100{,}000$ Five factors of the base 10 number

EXAMPLES: Rewrite the following using exponent notation.

1. 1,000,000,000,000,000

   10^{15} since there are 15 zeros which equate to 15 factors of 10

2. 10,000,000

   10^7 since there are 7 zeros which equate to 7 factors of 10

3. 100,000,000

   10^8 since there are 8 zeros which equate to 8 factors of 10

EXAMPLES: Rewrite the following by removing the exponent notation.

1. 10^3

   $10 \times 10 \times 10 = 1{,}000$ Notice the 3 zeros for the three factors of 10

2. 10^9

   $10 \times 10 \times 10 \times 10 \times 10 \times 10 \times 10 \times 10 \times 10 = 1{,}000{,}000{,}000$
   Notice the 9 zeros for the nine factors of 10

3. 10^4

$10 \times 10 \times 10 \times 10 = 10{,}000$ Notice the 4 zeros for the four factors of 10

Using this new exponent notation, we can rewrite the place-values as follows:

$10 = 10^1$, $100 = 10^2$, $1{,}000 = 10^3$, $10{,}000 = 10^4$, $100{,}000 = 10^5$, etc.

This means we can rewrite the place-value chart as follows:

…	One Hundred Thousands	Ten Thousands	Thousands	Hundreds	Tens	Ones	Decimal Point
…	10^5	10^4	10^3	10^2	10^1	1	.

Note: An alternative form of writing the exponent is using the carat symbol on the computer or calculator, ^, instead of as a superscript. This is typically the case when using a calculator. The notation looks like this.

10^1 = 10^1, 10^2 = 10^2, 10^3 = 10^3, 10^4 = 10^4, 10^5 = 10^5, etc.

Using exponential notation, we can now write our numbers in a more condensed expanded form as shown in the following examples.

EXAMPLES: Rewrite the following numbers in expanded form using exponent notation.

1. 34,365

$34,365 = 30{,}000 + 4{,}000 + 300 + 60 + 5$

$= 3 \times 10{,}000 + 4 \times 1{,}000 + 3 \times 100 + 6 \times 10 + 5 \times 1$

$= 3 \times 10^4 + 4 \times 10^3 + 3 \times 10^2 + 6 \times 10^1 + 5 \times 1$

2. 298,304

$298,304 = 200{,}000 + 90{,}000 + 8{,}000 + 300 + 4$

$= 2 \times 100{,}000 + 9 \times 10{,}000 + 8 \times 1{,}000 + 3 \times 100 + 0 \times 10 + 4 \times 1$

$= 2 \times 10^5 + 9 \times 10^4 + 8 \times 10^3 + 3 \times 10^2 + 0 \times 10^1 + 4 \times 1$

3. 6,930,496

$6{,}930{,}496 = 6{,}000{,}000 + 900{,}000 + 30{,}000 + 400 + 90 + 6$

$= 6 \times 1{,}000{,}000 + 9 \times 100{,}000 + 3 \times 10{,}000 + 0 \times 1{,}000 + 4 \times 100 + 9 \times 10 + 6 \times 1$

$= 6 \times 10^6 + 9 \times 10^5 + 3 \times 10^4 + 0 \times 10^3 + 4 \times 10^2 + 9 \times 10^1 + 6 \times 1$

EXERCISES 1.5

Exponents and Powers of Ten

Rewrite the expressions using exponents

1. $10 \cdot 10 \cdot 10 \cdot 10 \cdot 10 \cdot 10 \cdot 10$
2. $(10)(10)(10)$
3. $10 \times 10 \times 10 \times 10 \times 10 \times 10$
4. $(10)(10)(10)(10)(10)(10)(10)$
5. $10 \cdot 10 \cdot 10 \cdot 10 \cdot 10 \cdot 10 \cdot 10 \cdot 10 \cdot 10 \cdot 10 \cdot 10$
6. $10 \times 10 \times 10 \times 10$

Rewrite without exponents, and multiply out

7. 10^3
8. 10^5
9. $^{-}$10^4
10. 10^9
11. 10^5
12. $^{-}10^{6}$

Rewrite in expanded exponent notation form

13. 73
14. 56
15. 147
16. 289
17. 5,024
18. 8,290
19. 13,390
20. 130,450
21. 5, 689,003
22. 21,000
23. 200,395
24. 70,402
25. 975,389
26. 8,381
27. 2,450,000,000
28. 150,003,001,202

In the next section we extend this idea to other bases and introduce some rules for simplifying more complicated terms with exponents.

1.6 POSITIVE EXPONENTS IN GENERAL

In addition to multiplying powers of 10 a number of times, we will often have to multiply other numbers by themselves several times, such as 3 multiplied by itself 4 times, or 5 factors of 3. Instead of writing $3 \cdot 3 \cdot 3 \cdot 3 \cdot 3$ all the time, we can again use the exponent or power operator.

Thus, instead of $3 \cdot 3 \cdot 3 \cdot 3 \cdot 3$, we would write 3^5 in a mathematical expression. As we have already shown, the exponent operator is simply a shorthand way of writing a number multiplied by itself a fixed number of times. The number we are multiplying is called the base of the exponent operator and the number of times we are multiplying the same number is called the exponent.

EXAMPLES:

1. $2^4 = 2 \cdot 2 \cdot 2 \cdot 2 = 16$, the base is 2 and the exponent is 4, read as "two to the fourth" or "two to the fourth power"
2. $3^6 = 3 \cdot 3 \cdot 3 \cdot 3 \cdot 3 \cdot 3 = 729$, the base is 3 and the exponent is 6, read as "three to the sixth" or "three to the sixth power"
3. $5^3 = 5 \cdot 5 \cdot 5 = 125$, the base is 5 and the exponent is 3, read as "five to the third" or "five cubed"
4. $8^2 = 8 \cdot 8 = 64$, the base is 8 and the exponent is 2, read as "eight to the second" or "eight squared"
5. $5 \cdot 5 \cdot 5 \cdot 5 \cdot 5 \cdot 5 \cdot 5 \cdot 5 \cdot 5 = 5^9$
6. $^{-}6 \cdot {}^{-}6 \cdot {}^{-}6 \cdot {}^{-}6 \cdot {}^{-}6 \cdot {}^{-}6 \cdot {}^{-}6 = {}^{-}6^7$ or $({}^{-}6)^7$

*Note: When the superscript notation is used, you do not have to write parentheses around the negative number. However, in future courses, if the superscript notation is not used, the answers may not be the same. It all depends upon whether or not the exponent is an even or an odd number. This can cause a lot of confusion that we are avoiding for now, but it may come up again in a later course. Again, the symbol "–" is subtraction or "take the negative of" (an action or verb) whereas the symbol " ⁻ " means we are talking about a negative number (a name or noun). We should also make you aware that the symbol/button "(–)" on the calculator does not mean you are making the quantity into a negative number, instead this button simply performs the operation of "take the negative of what follows." It is just another action, and not a noun.

7. $-5^2 = -5 \cdot 5 = -25$, Notice how this is really "take the negative of 5 squared" and not "negative 5 squared." This is what was discussed in the note of part f above.

8. $\left(\frac{2}{3}\right)^4 = \left(\frac{2}{3}\right) \cdot \left(\frac{2}{3}\right) \cdot \left(\frac{2}{3}\right) \cdot \left(\frac{2}{3}\right) = \frac{2 \cdot 2 \cdot 2 \cdot 2}{3 \cdot 3 \cdot 3 \cdot 3} = \frac{16}{81}$

EXERCISES 1.6

Positive Exponents

Rewrite the expressions using exponents

1. $3 \cdot 3 \cdot 3 \cdot 3 \cdot 3 \cdot 3 \cdot 3$
2. $7 \cdot 7 \cdot 7$
3. $^-2 \cdot {^-2} \cdot {^-2} \cdot {^-2} \cdot {^-2} \cdot {^-2} \cdot {^-2}$
4. $^-3 \cdot {^-3} \cdot {^-3} \cdot {^-3} \cdot {^-3} \cdot {^-3}$
5. $\left(\frac{3}{5}\right) \cdot \left(\frac{3}{5}\right) \cdot \left(\frac{3}{5}\right)$
6. $\left(\frac{1}{2}\right) \cdot \left(\frac{1}{2}\right) \cdot \left(\frac{1}{2}\right) \cdot \left(\frac{1}{2}\right) \cdot \left(\frac{1}{2}\right)$
7. $\left(\frac{2}{3}\right) \cdot \left(\frac{2}{3}\right) \cdot \left(\frac{2}{3}\right) \cdot \left(\frac{2}{3}\right) \cdot \left(\frac{2}{3}\right) \cdot \left(\frac{2}{3}\right) \cdot \left(\frac{2}{3}\right)$

Rewrite without exponents, and multiply out

8. 4^3
9. 2^5
10. $(^-3)^3$
11. $(^-2)^4$
12. -3^4
13. -5^2
14. $^-6^2$
15. $^-4^2$
16. $\left(\frac{2}{3}\right)^4$
17. $\left(\frac{1}{3}\right)^2$
18. $\left(\frac{3}{4}\right)^3$
19. -4^3
20. $^-3^4$
21. $\left(\frac{1}{4}\right)^3$
22. 1^8
23. -1^{12}
24. $\left(\frac{5}{2}\right)^4$
25. $\left(-\frac{2}{7}\right)^2$

Chapter 1 Practice Test

Write each number in words, expanded form, using a place–value chart, and on an abacus.

1. 29
2. 36
3. 125

Given each of the following numbers, identify the place value of the underlined digit.

4. $2{,}32\underline{6}{,}900$
5. $\underline{6}5{,}280$

In each of the following problems, add or subtract using the expanded form and the standard algorithm.

6. $33+29$
7. $129-36$
8. $322+789$
9. $2{,}435-1{,}854$

In each of the following problems, add or subtract using the tape diagram.

10. $33+22$
11. $34-12$

12. Joan purchased 349 markers. Sue purchased 78 markers. How many markers did the two of them purchase together?

13. Kayla has 94 marbles. Michael has 18 less marbles that Kayla. How many marbles does Michael have?

Solve the following multiplication problems with a tape diagram, place–value chart with disks, expanded notation, partial products, and the standard algorithm.

14. 2×12
15. 3×15
16. 5×33
17. There are 22 cookies in each box of cookies. Denise has 6 boxes of cookies. How many cookies does Denise have?

Solve using an area model.

18. $40 \div 7$
19. $29 \div 6$

Solve the following division problems. Use the long division algorithm and/or draw tape diagrams to help illustrate what you solved. If there is a remainder, shade in a small portion of the tape diagram to represent that portion of the whole.

20. $33 \div 5$
21. $28 \div 4$
22. $74 \div 12$
23. Justin bought a package of 240 party favors to give to the guests at a friends birthday party. He calculated that he could give 12 party favors to each guest. How many guests is he expecting?

Rewrite using exponent notation

24. $10 \times 10 \times 10 \times 10$
25. $(10)(10)(10)(10)(10)$

Simplify by writing without exponents

26. 10^5
27. -10^4
28. 10^3

Write in expanded form using exponent notation.

29. 129,709
30. 16,238

Rewrite the expressions using exponents

31. $4 \cdot 4 \cdot 4 \cdot 4 \cdot 4$
32. $^{-}7 \cdot {}^{-}7 \cdot {}^{-}7 \cdot {}^{-}7$
33. $\left(\frac{3}{5}\right) \cdot \left(\frac{3}{5}\right) \cdot \left(\frac{3}{5}\right)$

Rewrite without exponents, and multiply out

34. 5^2
35. $({}^{-}2)^4$
36. -2^4
37. $\left(\frac{2}{3}\right)^2$

CHAPTER 2

Geometry, Measurement and Units

In this chapter we focus on using whole numbers to measure quantities. Whereas in Chapter 1 the focus was primarily on using whole numbers to count, in Chapter 2 we look at whole numbers as a way to measure. The ideal subject area to "measure" in is that of geometry. Thus, in this chapter we introduce the basics of geometry and geometrical units, and show how to use whole numbers to measure some basic attributes of geometric figures.

2.1 GEOMETRIC UNITS OF MEASURE

The most fundamental attribute we measure in geometry is a length. You might take out a tape measure and measure the length of a wall, or the length of a sidewalk, or the length of your body, also called your height, which is just another word for length. Typical units for length in the English system of measurement are inches (in), feet (ft), and miles (mi). In the metric system you might measure a length in meters (m), centimeters (cm), millimeters (mm), or kilometers (km). All measures of length are said to be one-dimensional, since they only require one unit of length to measure their value.

If we multiply two lengths together, then we have an area. Or, $\text{Area} = \text{Length} \times \text{Length}$. Typical units of area in the English system are square-inches (in^2), or square-feet (ft^2), or square-miles (mi^2). Similarly in the metric system you would have square-meters (m^2), square-centimeters (cm^2), square-millimeters (mm^2), or square-kilometers (km^2). An area measure is said to be two-dimensional, since it requires two lengths to measure its value.

If we multiply an area by a length, or if we multiply three lengths together, we have a volume. Or, $\text{Volume} = \text{Length} \times \text{Area} = \text{Length} \times \text{Length} \times \text{length}$. Typical units of volume in the English system are cubic-inches (in^3), or cubic-feet (ft^3). In the metric system you'll see cubic-meters (m^3), cubic-centimeters (cm^3), cubic-millimeters (mm^3) used a lot. A volume measure is said to be three-dimensional, since it requires three lengths to measure its value. Volumes also have another common form of measure. A volume is also a capacity, or how much "stuff" you can hold. Sometimes we have fluids, and it may not be possible to easily measure the three lengths of an amount of fluid. So when we have fluids we usually use different volume measures. In the English system we use ounces (oz), pints (pt), quarts (qt), or gallons (gal). In the metric system we have Liters (L) and milliliters (ml).

You should become familiar with the common units of measure in both the English and the metric systems. You should also be able to distinguish between one, two and three-dimensional objects.

EXERCISES 2.1

Determine the type of measure (length, area, or volume) and the dimension of each object described below, and then give a possible example of units to measure it.

1. The circumference of a circle
2. A bag of cement
3. The surface of a desk
4. The length of a string
5. A bucket full of water
6. A sheet for a bed
7. The distance to the sun
8. The size of a planet
9. Cargo space of an SUV
10. Laptop screen size (as advertised)
11. Waist measurement for jeans
12. Fuel tank capacity of a school bus
13. Supermarket receipt
14. Kitchen cabinet door
15. Length of Olympic size swimming pool
16. The capacity of a washing machine
17. The four walls of a room
18. Topsoil dumped from a truck
19. A path around a lake
20. A pool cover
21. A bottle of soda
22. The interior space of an automobile

2.2 BASIC GEOMETRY

In this section we are going to learn how to calculate the lengths and areas associated with some basic planar figures, as well as the volume of a very basic three-dimensional figure.

Aside from the length of a side of an object, another common length we are asked to find is the perimeter. The **perimeter** is the length along the outside or boundary of the object. You can visualize this as taking a flexible tape measure and wrapping it around the outside of the object, and then pulling the tape measure straight and reading how far around it is. See the figure below showing the idea of a perimeter of a planar object called a pentagon.

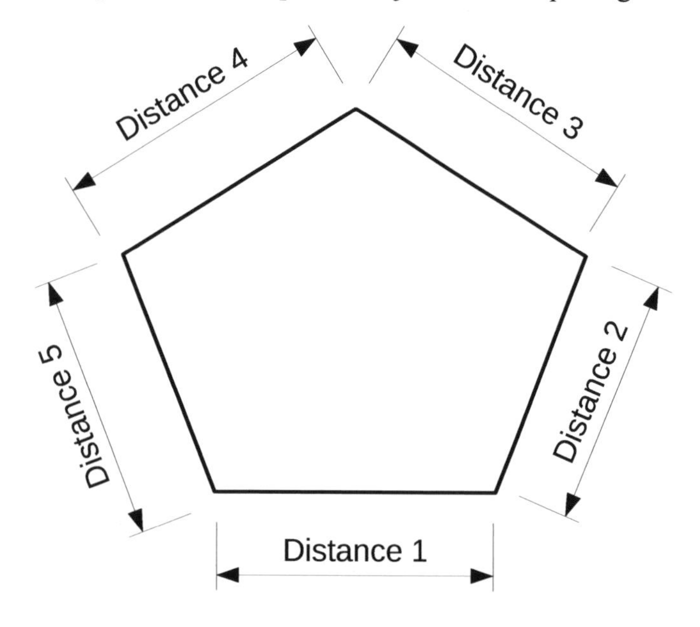

Perimeter is the sum of all the distances around the object. Its units are always a length.

Another way to think of finding or measuring the perimeter, is to imagine using unit length straight edges, and laying them end to end around the boundary of the object, and counting how many of these lengths we have. That is the measure of the perimeter. The unit length we use could be any measure we want, such as: inch, foot, mm, cm etc.

The other common quantity we calculate is the **area** of an object. In area calculations, we count the number of squares, that are of a fixed unit size, that will completely cover the object. This is why units of area are units squared. Sometimes we have an exact fit, although other times we cannot fit an entire square into our object, and can only use part of it. In this case, we see decimal or fractional areas. See the figure below.

11	12	13	14	15
6	7	8	9	10
1	2	3	4	5

Area counts the number of squares it takes to completely cover an object

It takes 15 squares (square units) to cover this rectangle so its area is 15 units ^{2}

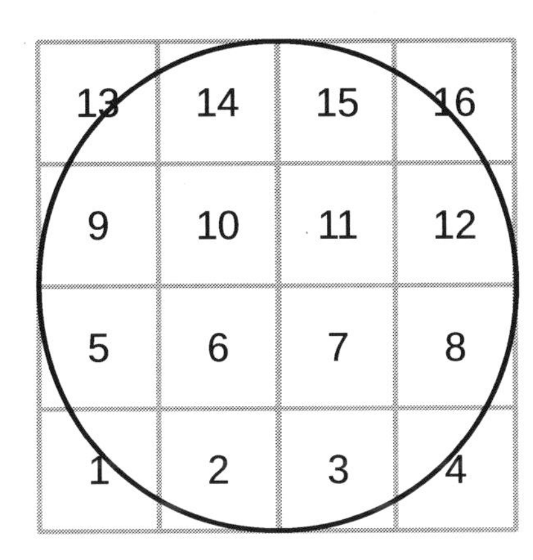

It takes less than 16 whole squares to cover this circle. In fact it takes a little over 12 and one half squares, so its area is around 12.5 units ^{2}

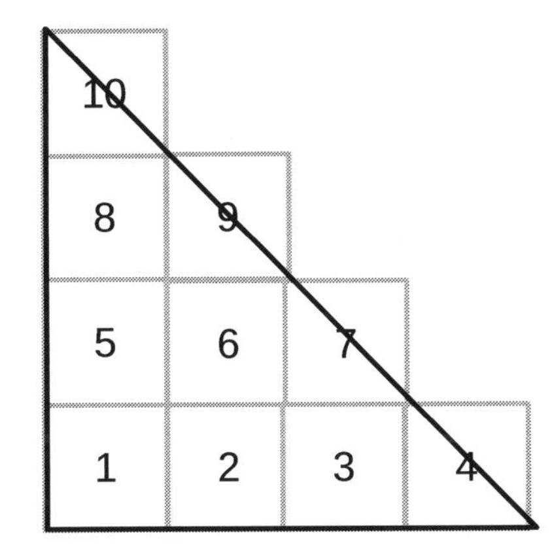

It takes 6 whole squares and 4 half squares for a total of 8 squares to cover this triangle so its area is 8 units ^{2}

The last quantity that we are interested in calculating is the **volume**. The volume counts how many cubes of a fixed size it takes to completely fill our solid, or three-dimensional object. See the figure below to get the general idea.

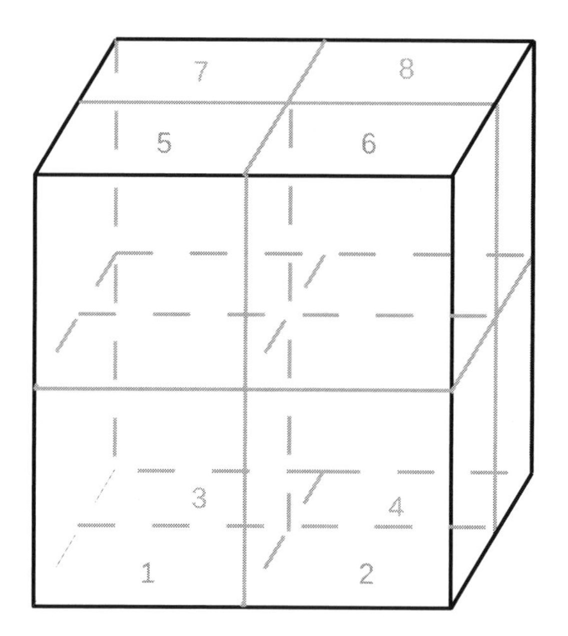

It takes 8 whole cubes to fill this larger cube, so its volume is 8 units ^{3}

We could go about "covering" quantities we are asked to measure with either lines, squares or cubes, but that would be fairly difficult and time consuming, and may not be accurate enough for some of our needs. This is the power of geometry. It provides formulas that give us the same results as above, but much more easily, and our answers can be as precise as we like.

Some of the more basic formulas are given below:

Object	**Figure**	**Formulas**
Square	L L	Perimeter = $4 \times L$ Area = L^2
Rectangle	L W	Perimeter = $2L + 2W$ Area = $L \times W$

Object	**Figure**	**Formulas**
Triangle	A B C	Perimeter = A + B + C
Prism or Rectangular Box	L W H	Volume = $L \times W \times H$

Later on in the text, when we introduce other types of numbers such as fractions and decimals we'll show some other common geometric formulas. For now, however, let's calculate these quantities for a few simple examples from the formulas above.

EXAMPLE 1: Find the perimeter and area of the square below.

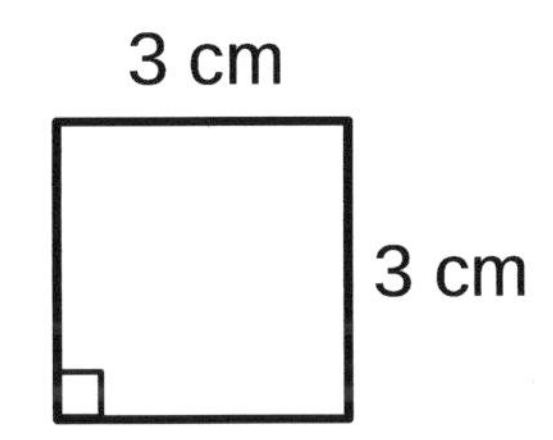

$$\text{Perimeter} = 4 \times \text{L} = 4 \times 3\,\text{cm} = 12\,\text{cm}$$

$$\text{Area} = \text{L}^2 = (3\,\text{cm})^2 = (3\,\text{cm}) \times (3\,\text{cm}) = 3 \times 3 \times \text{cm} \times \text{cm} = 9\,\text{cm}^2$$

EXAMPLE 2: Find the perimeter and area of the rectangle below.

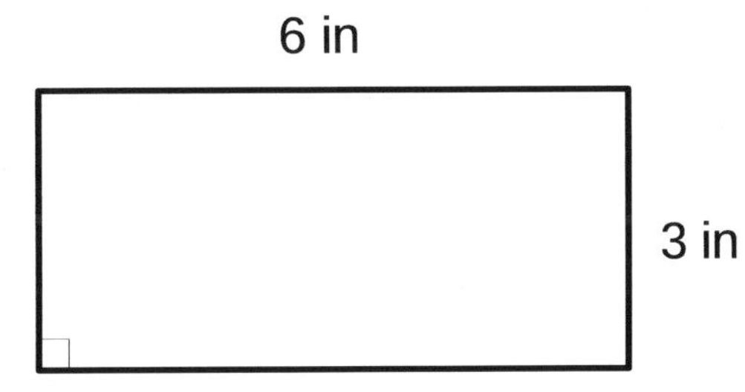

$$\text{Perimeter} = 2 \times \text{L} + 2 \times \text{W} = 2 \times 6\,\text{in} + 2 \times 3\,\text{in} = 12\,\text{in} + 6\,\text{in} = 18\,\text{in}$$

$$\text{Area} = \text{Length} \times \text{Width} = 6\,\text{in} \times 3\,\text{in} = 6 \times 3 \times \text{in} \times \text{in} = 18\,\text{in}^2$$

EXAMPLE 3: Find the perimeter of the triangle below.

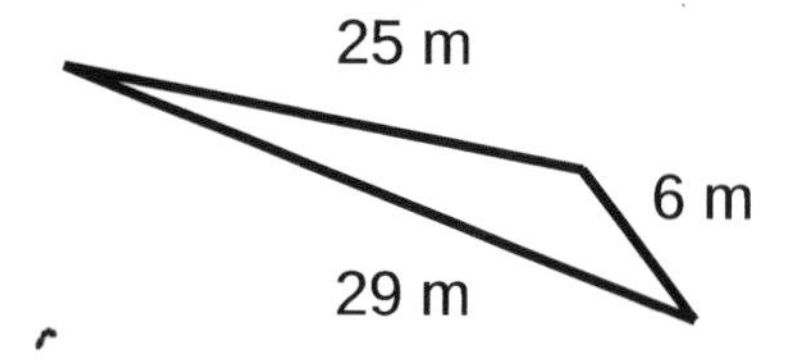

$$\text{Perimeter} = A+B+C = 25\text{m}+29\text{m}+6\text{m} = 60\text{m}$$

EXAMPLE 4: Find the volume of a rectangular box with width 15cm, length 20cm, and height 10cm.

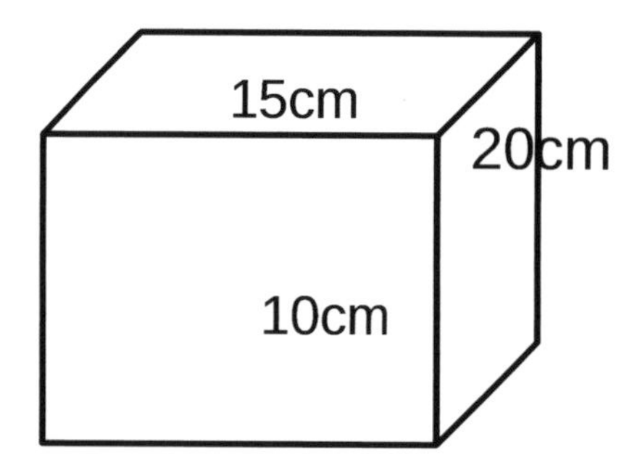

$$\begin{aligned} \text{Volume} &= L \times W \times H \\ &= 20\,\text{cm} \times 15\,\text{cm} \times 10\,\text{cm} \\ &= 3000\ \text{cm}^3 \end{aligned}$$

Since this is a three-dimensional object, the units are cubic-centimeters.

EXAMPLE 4: Find the volume of the cube shown below.

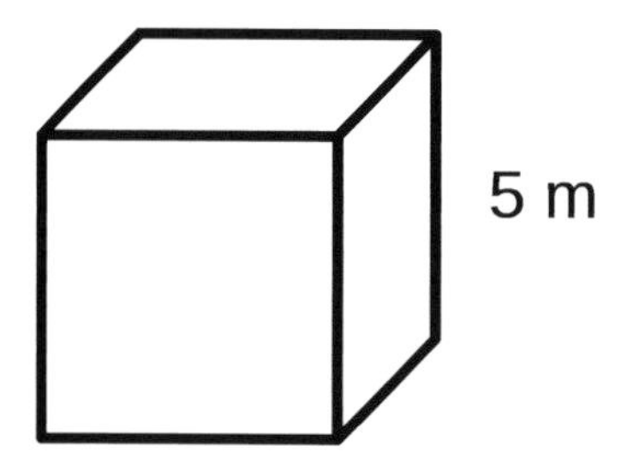

We should point out that a cube is simply a rectangular box, but with all the sides (Length, Width and Height) having the same length.

$$\begin{aligned} \text{Volume} &= L \times W \times H \\ &= 5\text{m} \times 5\text{m} \times 5\text{m} = 125\ \text{m}^3 \end{aligned}$$

Since this is a three-dimensional object, the units are cubic-meters.

EXERCISES 2.2

1. Find the perimeter and area of each of the following figures:

a) square

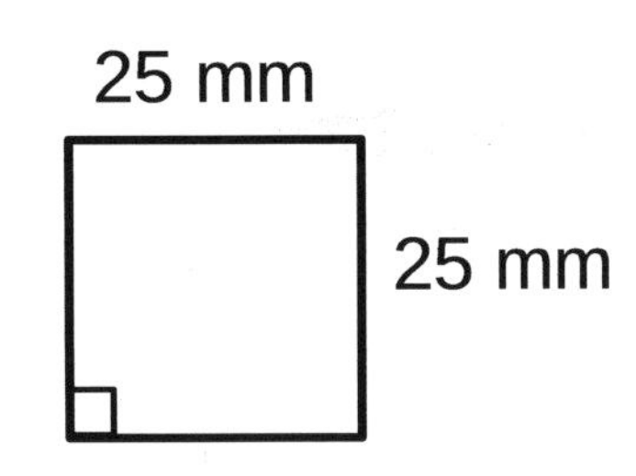

b) rectangle

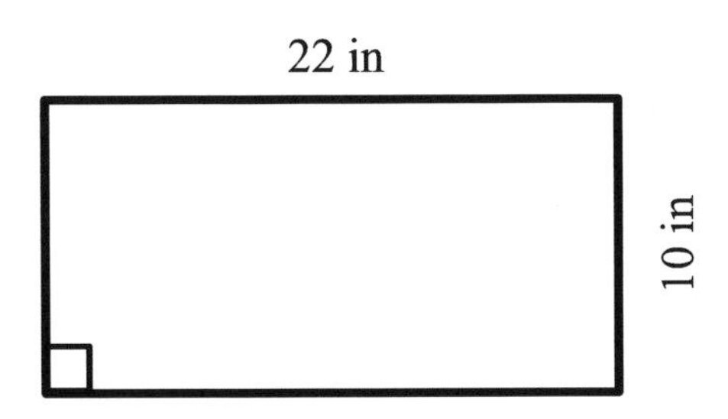

c) triangle

Perimeter only

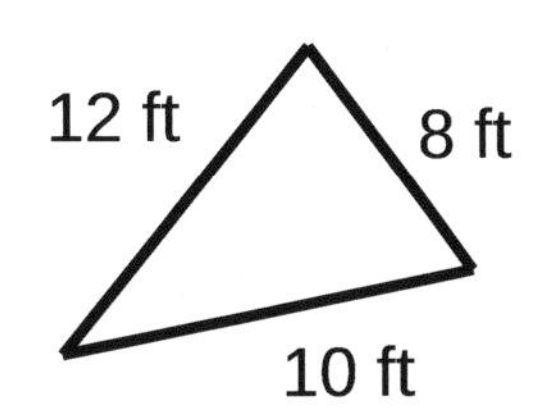

2. Find the volume of a cube given the length of one side.

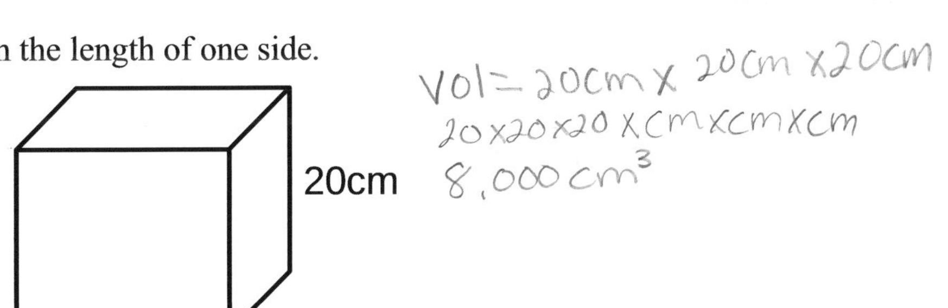

3. Find the volume of a rectangular box, given the sides as shown.

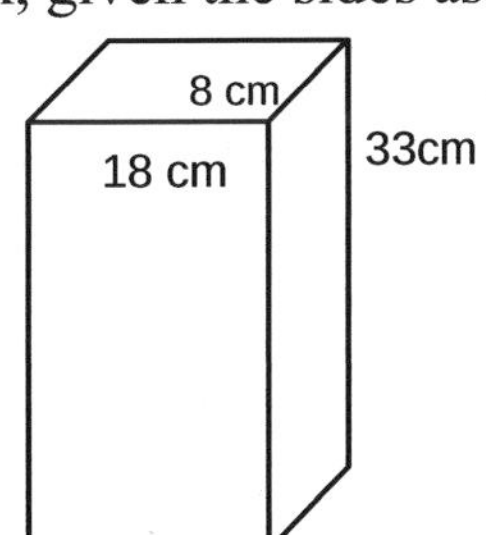

4. Find the volume of a rectangular solid that measures 15 inches in length by 18 inches in width, by 6 inches in height.
5. Find the perimeter of a triangle that measures 9 cm by 10 cm by 17 cm.
6. Find the perimeter of a triangle that measures 2 km by 3 km by 4 km.
7. Find the area of a rectangle that measures 25 cm in length by 10 cm in width.
8. Find the area of a rectangle that measures 4 m in length by 3 m in width.
9. Find the volume of a rectangular solid that measures 3 cm by 2 cm by 5 cm.
10. Find the volume of a cube whose side measures 4 inches.
11. Find the volume of a cube whose side measures 6 feet.

2.3 MORE ADVANCED GEOMETRIC APPLICATIONS

In this section we include some more complicated applications from geometry and separate them according to their arithmetic operations. We then relate this back to what we covered in Chapter 1.

Addition and Subtraction

Adding and subtracting whole numbers is equivalent to adding or subtracting whole amounts of attributes, and as long as we are adding or subtracting the same type of unit the operations make sense and can be done. Consider the following examples:

EXAMPLE 1: Find the perimeter of a rectangle whose sides are 36 centimeters by 24 centimeters.

It is always best to draw a picture of what you are trying to find. Here is a picture of the object described above.

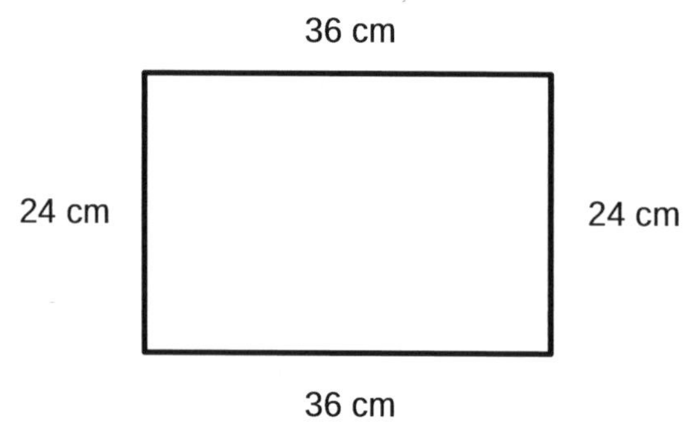

The perimeter is simply the sum of the lengths of the sides of the rectangle, or

$$36 \text{ cm} + 24 \text{ cm} + 36 \text{ cm} + 24 \text{ cm} = 120 \text{ cm}$$

Notice here that when we sum terms with units, the units not change. Summing centimeters means your final answer is in centimeters. Units will not change under addition or subtraction.

EXAMPLE 2: Find the perimeter of the room shown below.

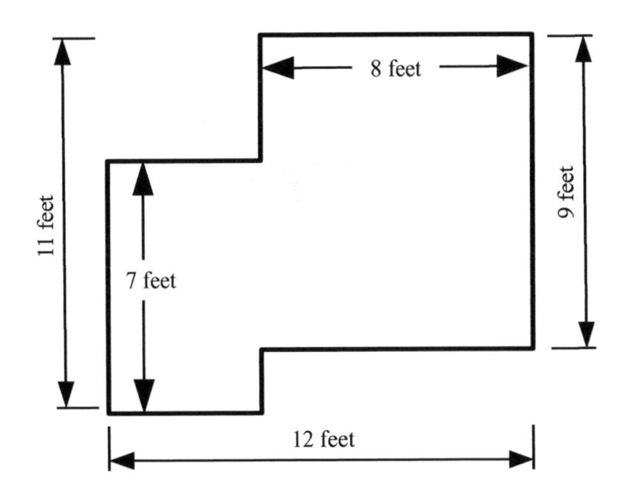

In this example we see that not all the lengths are given directly. If we show the direct lengths, along with the unspecified lengths; A, B, C, D, E, we have:

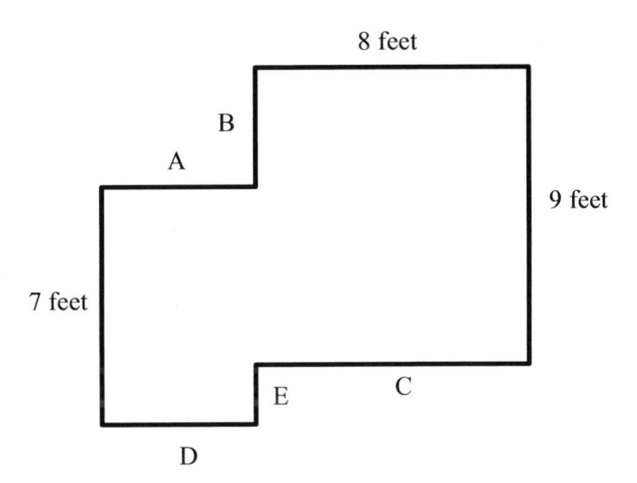

First we can see that side C is 8 feet long, since it is the same length as the side directly above it. Next we notice that sides A and D are equal in length due to the same reasoning, and their lengths can be found by subtracting 8 feet from 12 feet, or

$$A = D = 12 \text{ feet} - 8 \text{ feet} = 4 \text{ feet.}$$

In a similar way side B can be found by subtracting 7 feet from 11 feet, or

$$B = 11 \text{ feet} - 7 \text{ feet} = 4 \text{ feet.}$$

Finally, side

$$E = 11 \text{ feet} - 9 \text{ feet} = 2 \text{ feet.}$$

Replacing all the sides with their calculated lengths we have:

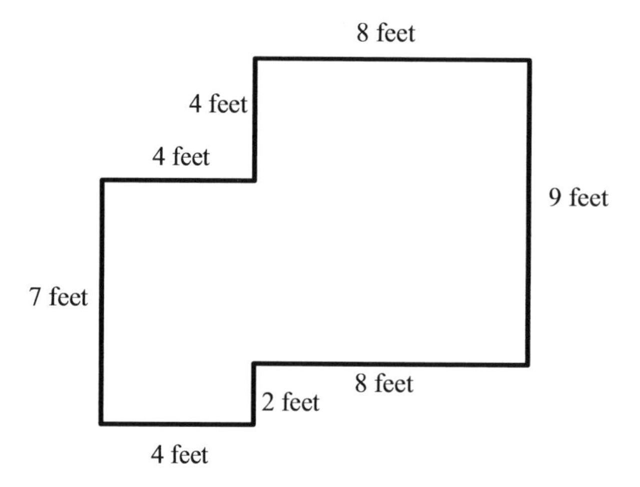

The perimeter is the sum of the lengths of all the sides:

Perimeter = 7 feet + 4 feet + 4 feet + 8 feet + 9 feet + 8 feet + 2 feet + 4 feet = 46 feet

Multiplication

In Chapter 1 we said that multiplication can be thought of as repeated additions. Now this is true if what we are calculating is simply multiple copies of a particular attribute. However, if we multiply two or more attributes together, the interpretation is different. We illustrate this distinction with some examples from geometry.

EXAMPLE 3: Find the perimeter of a square whose sides are 3 feet.

The sides of a square are the same length and there are 4 sides, or 4 copies of same length on a square. The answer is 4 repeated additions of 3 feet, or

$$\text{perimeter} = 4 \times 3 \text{ feet} = 12 \text{ feet}$$

EXAMPLE 4: Find the area of a rectangle that measures 5 meters by 16 meters.

For this problem it is best to illustrate with a picture. Here we draw a 5 meter by 16 meter rectangle and sub-divide it into square meters.

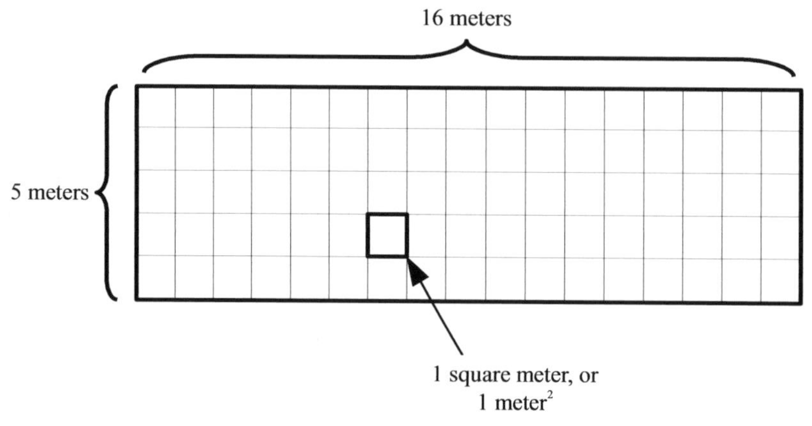

If we count the number of square meters we see that we have 80 of them, or

$$5 \text{ meters} \times 16 \text{ meters} = (5 \times 16)\,(\text{meters} \times \text{meters}) = 80 \text{ meters}^2$$

Now, we can also think of this as 5 repeated additions of 16, but now we are actually adding square meters, or areas, and NOT lengths! Thus, you can see when we have units, the earlier interpretation of multiplication now has to change. It is still repeated additions, but now it is repeated addition of the square units.

EXAMPLE 5: Find the volume of a rectangular solid that measures 5 centimeters by 3 centimeters by 2 centimeters.

We start with a picture to see what we are being asked to find.

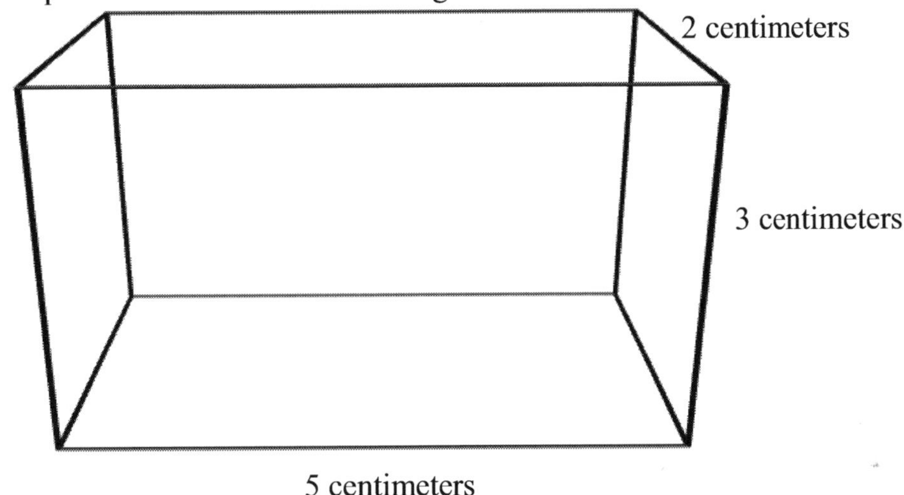

We are now multiplying three lengths. If we multiply two lengths we get an area. Multiplying three lengths together gives us a volume, which in this case are cubic centimeters. If we subdivide our object into cubic centimeters we have:

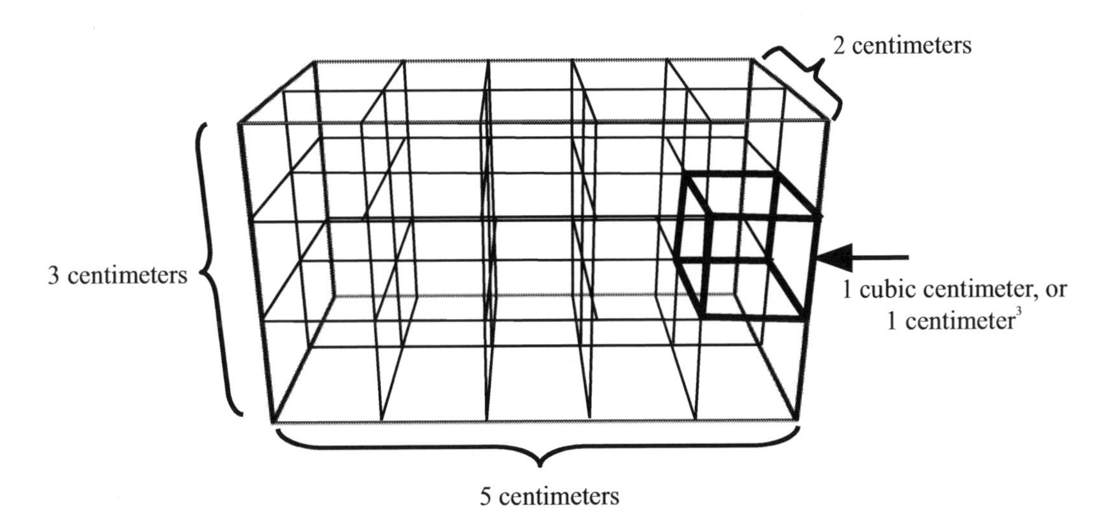

If we count the number of cubic centimeters we have we will find that there are 30 of them. So the answer is 30 centimeters^3, or

$$3 \text{ centimeters} \times 5 \text{ centimeters} \times 2 \text{ centimeters} = (3 \times 5 \times 2)(\text{centimeters} \times \text{centimeters} \times \text{centimeters})$$

$$= 30 \text{ centimeters}^3$$

Again, we can also think of this as either 3 repeated additions of 10 cubic centimeters, or 2 repeated additions of 15 cubic centimeters, or even 15 repeated additions of 2 cubic centimeters.

Division

Division is the inverse operation of multiplication. Thus, instead of finding the result of a fixed number of repeated additions, you are instead either looking for the number of repeated additions, or the fixed number you repeatedly add to get the final result. You are instead looking for the inverse result.

When we view division from a measurement perspective, as the amount of some attribute divided by a fixed number, or an amount of one attribute divided by the amount of another attribute, we have to be careful in interpreting the final result. The best way to illustrate this is through some examples.

EXAMPLE 6: Find the area of a 15 square feet rectangular region divided by 3, or stated in an equivalent way, divided into 3 separate regions.

Again, let's draw an image of what we are being asked to find.

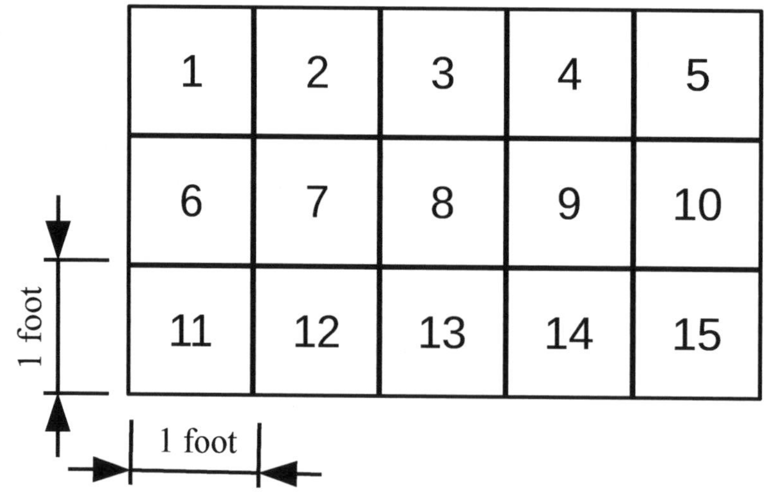

Here we have a 15 square foot area and we now divide it into 3 equal parts.

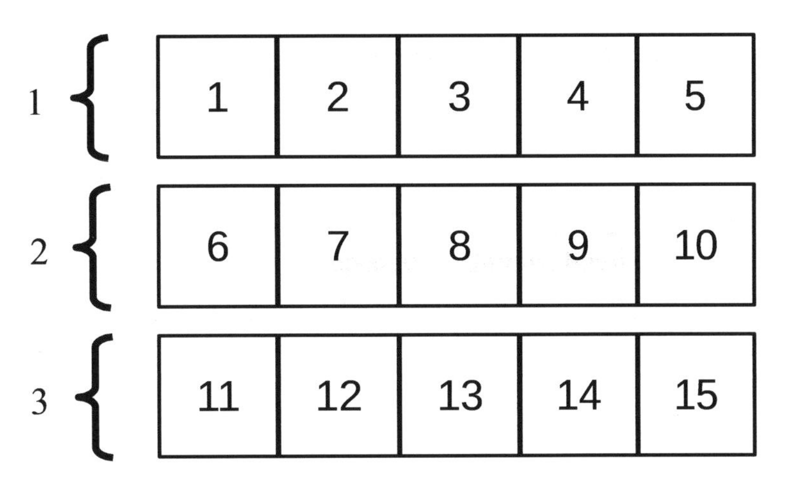

This leaves us with 5 square feet in each part so our answer is 5 square feet, i.e.

$$\frac{15\,\text{feet}^2}{3} = \frac{15}{5}\ \text{feet}^2 = 3\ \text{feet}^2$$

EXAMPLE 7: Divide the area of a 16 square meter rectangle, by a length of 8 meters.

Draw a picture:

In this example we are dividing an area (meter^2) by a length (meter). From the drawing we see that we have an area made up of 16 separate square each with an area of 1 square meter.

Now, when we do the calculation, this give us:

$$\frac{16\ \text{meter}^2}{8\,\text{meter}} = \frac{16\,\text{meter}\times\text{meter}}{8\,\text{meter}} = \frac{16}{8}\ \frac{\text{meter}\times\cancel{\text{meter}}}{\cancel{\text{meter}}} = 2\ \text{meter}$$

Notice how the meter units canceled, exactly like we would cancel two whole numbers that were the same value when dividing whole numbers in Chapter 1. This is always true with units. Like units cancel in division. The other interesting fact this tell us is that an *area* divided by a *length* is a *length*!

EXAMPLE 8: Divide the volume of a 42 cubic centimeter box divided by a 7 centimeter length.

In this example we'll follow the same reasoning from the previous example and write:

$$\frac{42\,\text{centimeter}^3}{7\,\text{centimeter}} = \frac{42\,\text{centimeter} \times \text{centimeter} \times \text{centimeter}}{7\,\text{centimeter}}$$

$$= \frac{42}{7}\,\frac{\text{centimeter} \times \text{centimeter} \times \cancel{\text{centimeter}}}{\cancel{\text{centimeter}}} = 6\ \text{centimeter}^2$$

This tells us that a *volume* divided by a *length* gives us an *area*.

EXERCISES 2.3

1. The perimeter of a rectangle is the distance around the entire rectangle. Find the perimeter if a rectangle that is 6 meters wide by 8 meters long.

2. The length of the shorter side of a rectangle is 5 feet shorter than the longer side. The longer side of a rectangle is 13 feet. How long is the shorter side of the rectangle?

3. Find the perimeter of the room below.

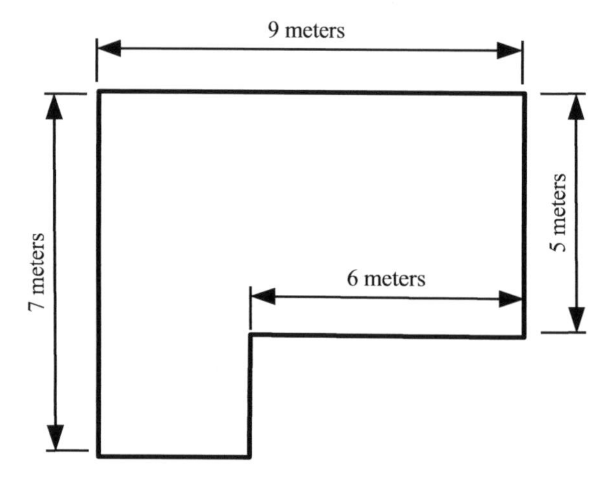

5. Find the area of a rectangle that is 12 meters by 6 meters. Also express your answer as repeated additions of a basic area unit.

6. Find the area of a square whose sides are all 6 inches long. Also express your answer as repeated additions of a basic area unit.

7. Find the volume of a rectangular solid whose sides are 8 centimeters, by 6 centimeters, by 5 centimeters. Also express your answer as repeated additions of a basic volume unit.

8. Find the volume of a rectangular solid that are 7 inches, by three inches, by 4 inches. Also express your answer as repeated additions of a basic volume unit.

9. Divide a 21 cubic inch volume by 7 inches.

10. Divide a 36 square centimeter rectangle by 4 centimeters.

11. Divide a 27 $meter^3$ rectangular solid by a 9 $meter^2$ area.

12. Divide a 38 foot pole by 19.

13. Divide a 125 square centimeter area by 25 square centimeters.

14. Divide 51 cubic inch volume by 3.

15. The area of a rectangular piece of land is 108 m^2. If the width of the land is 9 m, then what is the length.

Chapter 2 Practice Test

Determine the type of measure (length, area, or volume) and the dimension of each object described below, and then a possible example of units to measure it.

1. distance to the New York City from Selden,NY
2. cargo space of an airplane
3. surface of a desk

Answer the following:

4. Find the perimeter of a triangle that measures 3m by 4m by 5m.
5. Find the volume of a cube whose side measures 8 inches.
6. Find the area of a rectangle that measures 48 cm in length by 10 cm in width.
7. Find the volume of a rectangular solid that measures 4 cm by 3 cm by 7 cm.

Solve each of the following:

8. Find the perimeter of a rectangle that measures 22 cm in length and 18 cm in width.
9. Find the volume of a rectangular solid whose sides are 10 centimeters, by 8 centimeters, by 7 centimeters. Also express your answer as repeated additions of a basic volume unit.
10. Divide a 36 square centimeter rectangle by 6 centimeters.
11. Divide a 64 cubic centimeter volume by 4 centimeters.
12. Divide 54 cubic inch volume by 3.
13. Divide a 40 foot pole by 20.
14. The area of a rectangular desk top is 90 ft^2. If the width of the desk top is 9 ft, then what is the length of the desk top?

CHAPTER 3

Fractions – Parts of a Whole

3.1 INTRODUCTION

In this section we will show that fractions (parts of a whole) are numbers, just like the natural counting numbers, 1, 2, 3, … Numbers are essentially used to quantify things or phenomena we see in the world. The first numbers that came about were the natural or counting numbers. These numbers allowed us to answer questions like; How many? and Which sets are larger than other sets? When we count whole objects there are no in-between values. For example it does not make sense to talk about five and one-third people. There are either five or six people, there is no in-between value. However, as we stated early in Chapter 1, numbers need to be thought of more generally, as they can be used to quantify things that do have in-between values, such as, distances, time, weight, etc.. It is this more general idea of numbers that we explore further in this chapter.

A useful tool to understand the number concept, is the number line:

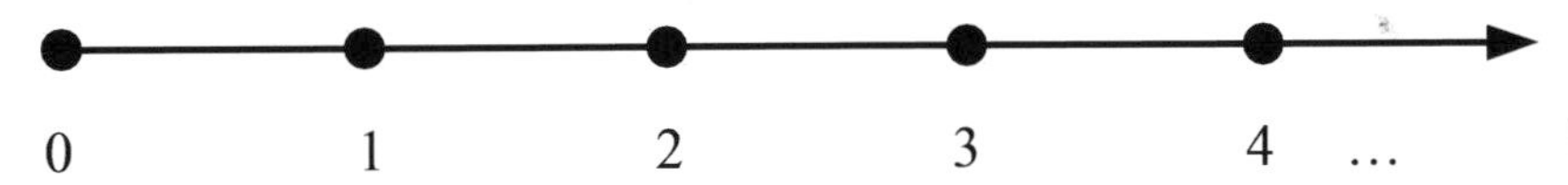

On this number line we have placed the counting numbers, along with zero, as points on the line. We have placed the numbers in order, with increasing values to the right. In this way we can think of the numbers as representing a distance from zero on the line. Thus, the numbers are not simply counting whole objects, but are now used to measure (count) distance. In this way we have extended the concept of number to quantify things that have values in-between the whole numbers.

To explore this further we focus our attention on a portion of the number line (a line segment) from zero to one:

0 1

From this unit line segment we will begin to develop the concept of fractions and its use as a number, just like the counting numbers.

We should point out that this is a fairly abstract concept. The Ancient Egyptians in 3,000 BCE were some of the earliest users of fractions, but did not consider them numbers. To the early Egyptians and most ancient civilizations, the only numbers were the counting numbers, and fractions such as:

$$\frac{1}{4}, \frac{1}{2}, \frac{1}{7}, \frac{3}{5}, \text{ etc.}$$

were simply two numbers used to represent parts of a whole, but were not thought to be a single numerical quantity themselves. One early civilization, the Babylonians (3,000 BCE), however, did view fractions as numbers.

3.2 UNIT FRACTIONS– Parts of a Whole

We now begin to develop the concept of the two numbers as written above as actually representing a single numerical quantity, something the Ancient Egyptians could not grasp.

We start with our unit line segment:

We now look to break this segment into equal parts, or sub-segments. The first thing we notice is that we can break this segment into many different equal parts. We could break it into two equal parts, three equal parts, four equal parts, etc. If we break our line segment into two equal parts we call each of the parts a half of the whole and represent this using the symbol that places the number 1 representing the whole line segment over the number 2 representing the number of equal parts were are breaking (or dividing) our unit line segment into, with a horizontal line between them:

$$\frac{1}{2}$$

This is the unit fraction called one half.

Using our line segment we can illustrate the two equal parts and their size represented by one-half:

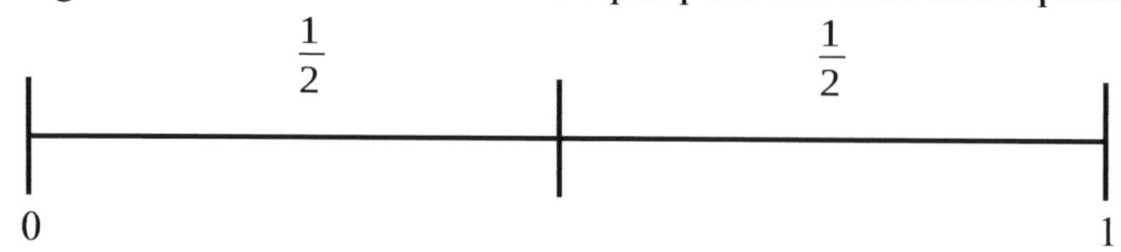

Breaking the line segment into three equal parts we have:

$$\frac{1}{3} \qquad \frac{1}{3} \qquad \frac{1}{3}$$

$$0 \qquad\qquad\qquad 1$$

With each equal part called a third.

Breaking into four equal parts we have:

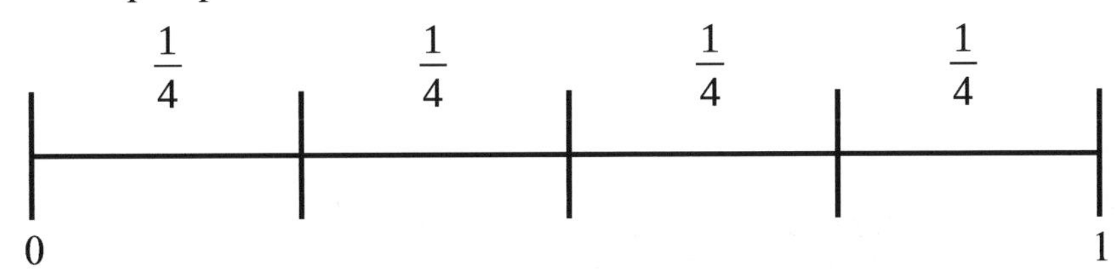

With each equal part being called a fourth, or a quarter, etc.

Each of the equal parts:

$$\frac{1}{2}, \frac{1}{3}, \frac{1}{4}, \text{ etc}.$$

are called unit fractions, since we are breaking a single part (the unit), into two or more equal parts.

Another useful way to understand fractions is by using an area approach in addition to the number line. Thus, along with the number line we can consider an area bar to represent a fraction. Sometimes we will refer to this as a "fraction bar."

Below are examples of representing one-half, one-third, and one-fourth, using a fraction bar:

$\frac{1}{2}$	$\frac{1}{2}$

$\frac{1}{3}$	$\frac{1}{3}$	$\frac{1}{3}$

$\frac{1}{4}$	$\frac{1}{4}$	$\frac{1}{4}$	$\frac{1}{4}$

At this point the fractions are not quite numbers, but instead illustrate a process where we take a whole object (the unit), and divide it into "n" equal parts to get the fraction,

$$\frac{1}{n}$$

To begin to develop the number concept for a fraction we see that any whole object can be decomposed (broken down into) into a sum of unit fractions. For example:

$$1 = \frac{1}{2} + \frac{1}{2}$$

We call this an addition sentence.

We can also visually represent this using the diagram below called the number bond.

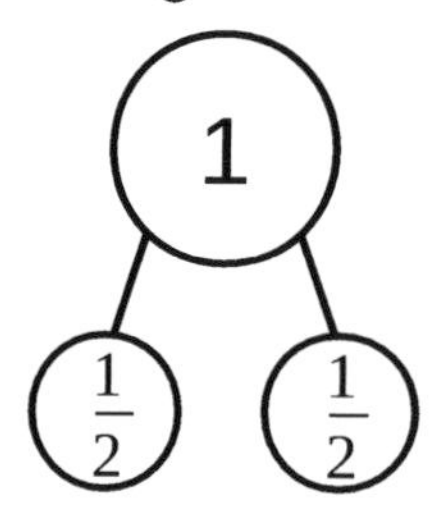

As another example, consider breaking one into thirds. We see that we can add it in two different ways. We could add all the thirds together in our addition sentence to get three thirds or one, or we could first add the two thirds together and then the one third and get the same result.

$$\begin{aligned} 1 &= \underbrace{\frac{1}{3} + \frac{1}{3}} + \frac{1}{3} \\ &= \frac{2}{3} + \frac{1}{3} \\ &= \frac{3}{3} \end{aligned}$$

We can also see this using a tape diagram which is actually equivalent to the fraction bar below.

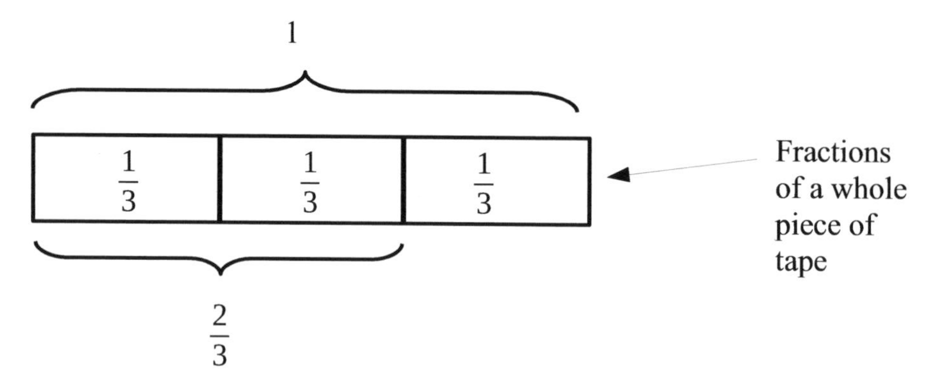

Furthermore, we see that two thirds can be thought of as adding two one thirds together or equivalently as multiplying 2 times a third, since multiplication as was defined earlier is simply repeated addition.

Written out in mathematical language, we call the first line an addition statement and the second line a multiplication statement.

$$\frac{2}{3} = \frac{1}{3} + \frac{1}{3} \qquad \text{Addition Statement}$$

$$\frac{2}{3} = 2 \times \frac{1}{3} \qquad \text{Multiplication Statement}$$

We can extend this process to consider fractions where the top number (the numerator) is greater than the bottom number (the denominator.) For example let's break up the fraction four thirds as shown below:

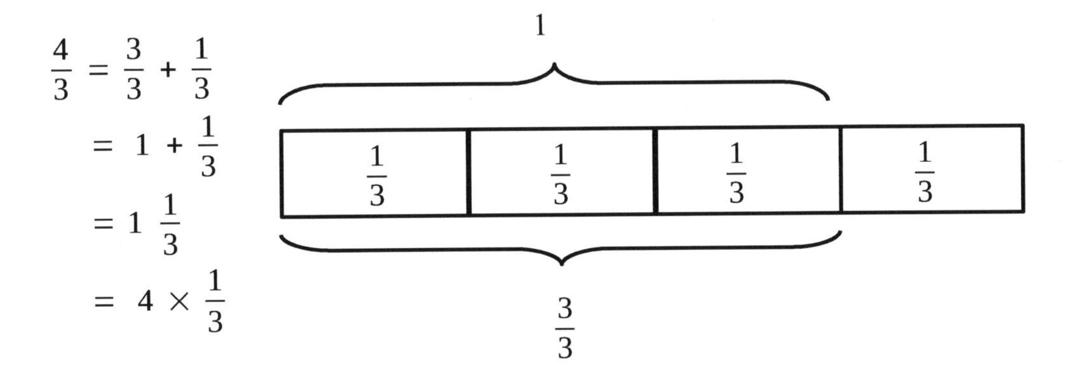

$$\frac{4}{3} = \frac{3}{3} + \frac{1}{3} = 1 + \frac{1}{3} = 1\frac{1}{3} = 4 \times \frac{1}{3}$$

As we see fractions are beginning to look more like numbers than we first thought. We can add them and we can multiply them by whole numbers.

EXAMPLE 1:

Decompose the fraction $\frac{7}{5}$ in at least three different ways, and also write using a multiplication statement.

Using the ideas from above we can represent $\frac{7}{5}$ in a few different ways, for example:

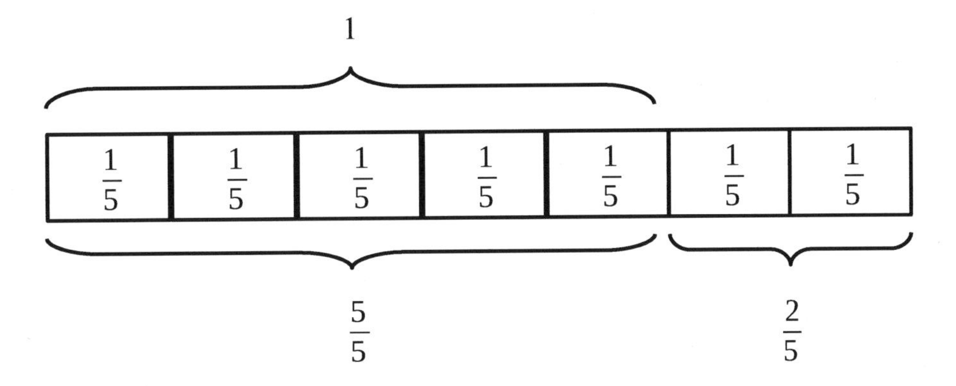

$$\frac{7}{5} = \frac{1}{5} + \frac{1}{5} + \frac{1}{5} + \frac{1}{5} + \frac{1}{5} + \frac{1}{5} + \frac{1}{5}$$

$$= \frac{6}{5} + \frac{1}{5}$$

$$= \frac{5}{5} + \frac{2}{5}$$

$$= 1 + \frac{2}{5}$$

$$= 1\frac{2}{5}$$

$$= 7 \times \frac{1}{5}$$

EXAMPLE 2:

Given the fraction $\frac{5}{6}$ use a number bond and an addition sentence to represent it using unit fractions. Then show an alternative way to write it. Finally represent its value using multiplication.

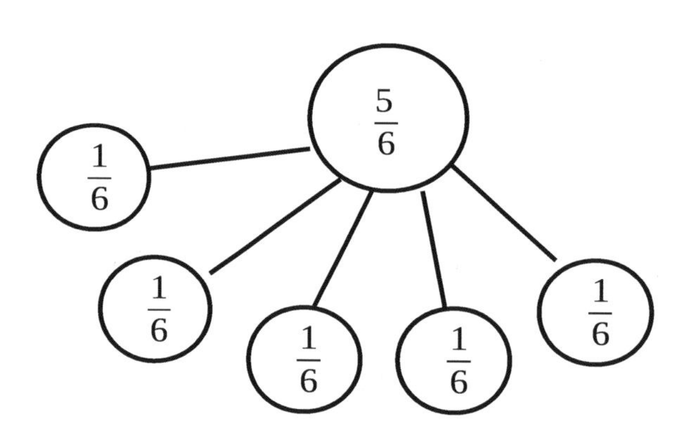

$$\frac{5}{6} = \frac{1}{6} + \frac{1}{6} + \frac{1}{6} + \frac{1}{6} + \frac{1}{6}$$

$$= \frac{4}{6} + \frac{1}{6}$$

$$= 5 \times \frac{1}{6}$$

EXERCISES 3.2

Unit Fractions

Represent the unit fractions using a tape diagram.

1. $\frac{1}{5}$
2. $\frac{1}{9}$
3. $\frac{1}{6}$
4. $\frac{1}{10}$

Represent the fractions using a tape diagram and a number bond. Then decompose the fractions into unit fractions, and then in at least 2 different ways. Write a suitable addition and multiplication statement for each.

5. $\frac{4}{5}$
6. $\frac{5}{6}$
7. $\frac{7}{9}$
8. $\frac{4}{7}$
9. $\frac{7}{4}$
10. $\frac{9}{5}$
11. $\frac{3}{4}$
12. $\frac{5}{8}$

3.3 Decompose Fractions Into Smaller Units

We just learned how to take a whole unit and decompose it into smaller unit fractions. We now extend the idea of decomposition to fractions themselves. We now ask the question; Can we decompose a unit fraction into smaller unit fractions?

EXAMPLE 1: Write $\frac{1}{3}$ as smaller unit fractions.

We could break $\frac{1}{3}$ into two smaller but equal pieces. As we see from the tape diagram and number bond, half of its size is equal to $\frac{1}{6}$, and it takes two $\frac{1}{6}$'s to equal $\frac{1}{3}$.

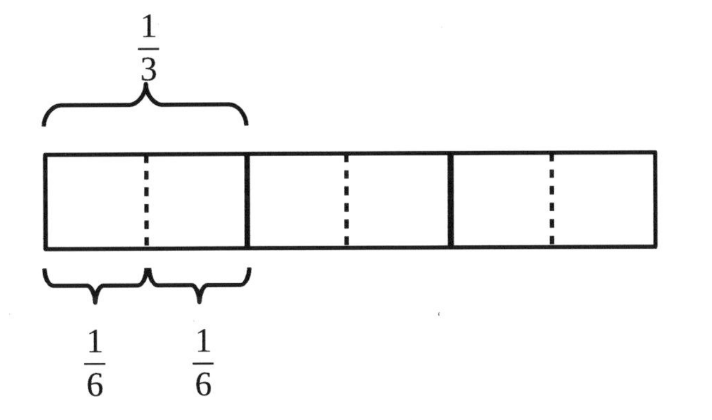

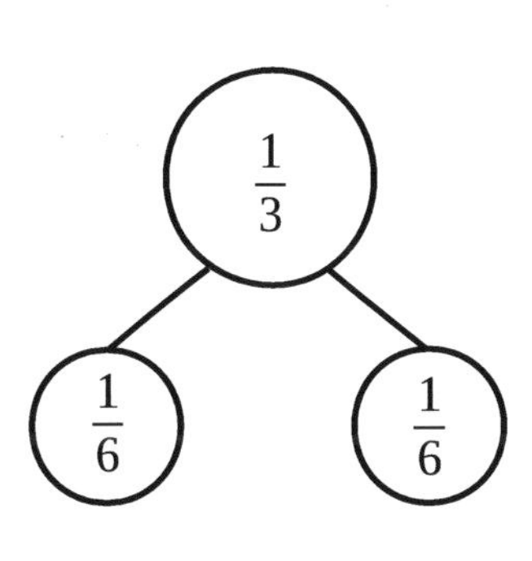

We can also write with number sentences, both addition and multiplication:

$$\frac{1}{3} = \frac{1}{6} + \frac{1}{6} = \frac{2}{6}$$

$$\frac{1}{3} = 2 \times \frac{1}{6} = \frac{2}{6}$$

Using the tape diagram, number bond, addition and multiplication sentences, we see that $\frac{1}{3}$ is equivalent to $\frac{2}{6}$.

We can take this even further.

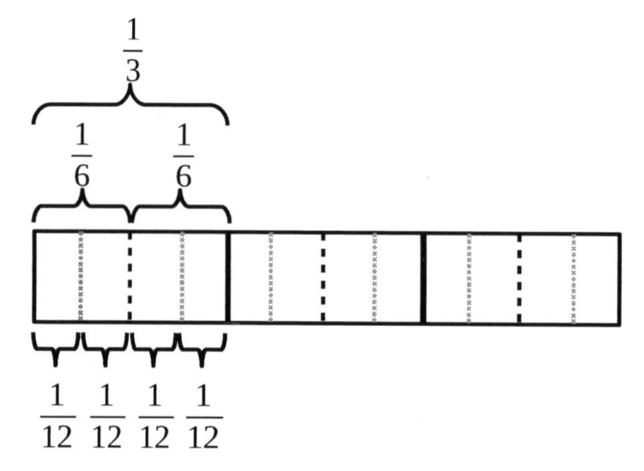

What does the figure above tell us about the relationships between $\frac{1}{3}$, $\frac{1}{6}$, and $\frac{1}{12}$?

We can write addition and multiplication sentences that describe this relationship.

$$\frac{1}{3} = \frac{1}{6} + \frac{1}{6} = \frac{2}{6}$$

$$\frac{1}{3} = 2 \times \frac{1}{6} = \frac{2}{6}$$

$$\frac{1}{3} = \frac{1}{12} + \frac{1}{12} + \frac{1}{12} + \frac{1}{12} = \frac{4}{12}$$

$$\frac{1}{3} = 4 \times \frac{1}{12} = \frac{4}{12}$$

EXAMPLE 2: Break $\frac{1}{5}$ into 3 equal sized pieces. Write addition and multiplication sentences that describe this relationship.

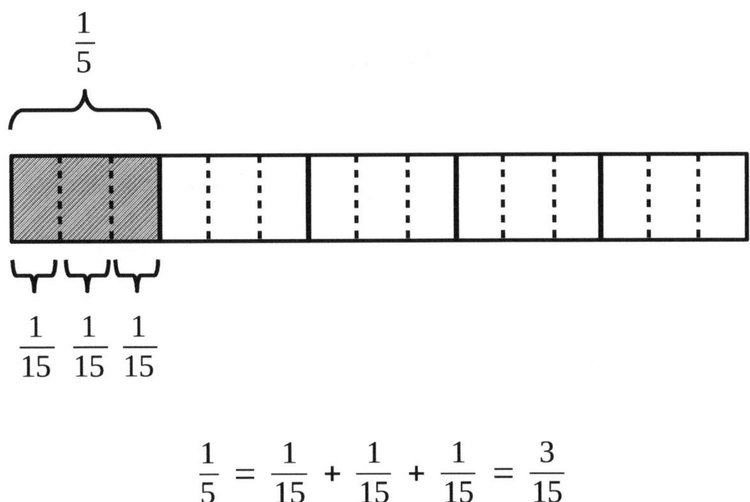

$$\frac{1}{5} = \frac{1}{15} + \frac{1}{15} + \frac{1}{15} = \frac{3}{15}$$

$$\frac{1}{5} = 3 \times \frac{1}{15} = \frac{3}{15}$$

EXAMPLE 3: Break $\frac{2}{6}$ into 4 equal sized pieces.

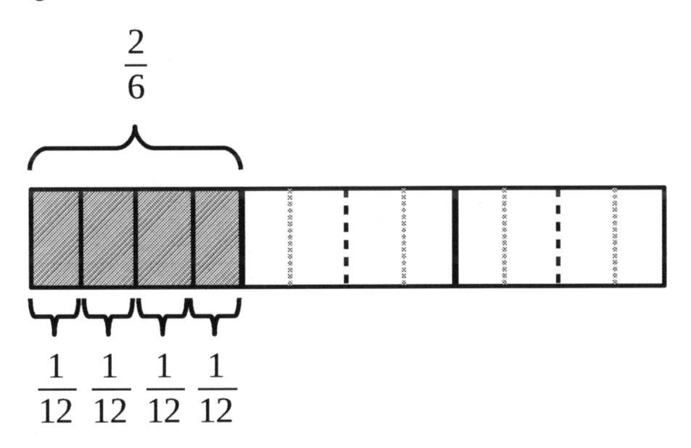

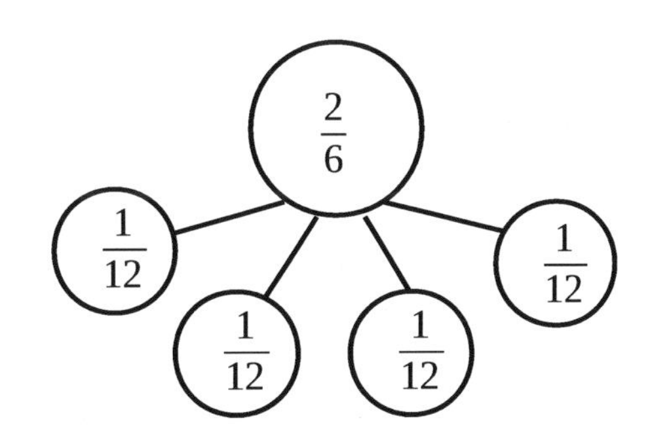

$$\frac{2}{6} = \left(\frac{1}{12} + \frac{1}{12}\right) + \left(\frac{1}{12} + \frac{1}{12}\right) = \frac{4}{12}$$

$$\frac{2}{6} = \left(2 \times \frac{1}{12}\right) + \left(2 \times \frac{1}{12}\right) = \frac{4}{12}$$

$$\frac{2}{6} = 4 \times \frac{1}{12} = \frac{4}{12}$$

This shows the relationship between $\frac{2}{6}$ and $\frac{4}{12}$

From this we see that $\frac{2}{6}$ and $\frac{4}{12}$ are equivalent. They have the same value. We call these two **Equivalent Fractions**. In the next section we look at many different ways to understand and find equivalent fractions.

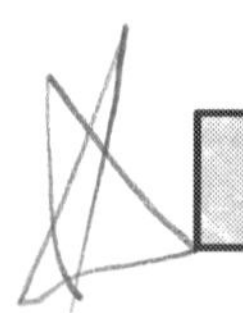

EXERCISES 3.3

Decompose Fractions Into Smaller Units

Break the given fractions into the given number of equal sized pieces. Show with a tape diagram. Then write addition and multiplication sentences that describe the relationship.

1. $\frac{1}{4}$ into 2 pieces
2. $\frac{1}{2}$ into 5 pieces
3. $\frac{1}{6}$ into 3 pieces
4. $\frac{1}{8}$ into 2 pieces
5. $\frac{4}{5}$ into 4 pieces
6. $\frac{7}{9}$ into 7 pieces
7. $\frac{2}{4}$ into 8 pieces
8. $\frac{4}{6}$ into 12 pieces
9. $\frac{5}{8}$ into 10 pieces.
10. $\frac{2}{3}$ into 4 pieces.
11. $\frac{2}{5}$ into 6 pieces.
12. $\frac{3}{10}$ into 9 pieces.

3.4 FRACTIONS AS NUMBERS AND EQUIVALENT FRACTIONS

We have shown that it is possible to count parts, just like we count wholes. For example, we can have one half, two halves, three halves, etc. Furthermore, in the same way we can count multiples of wholes, simply by multiplying by our counting numbers, such as:

$$1 = 1\times1,$$
$$2 = 1 + 1 = 2\times1,$$
$$3 = 1 + 1 + 1 = 3\times1,$$
etc.

we can do the same for unit fractions, and write the results as follows:

$$\frac{1}{2} = 1 \times \frac{1}{2},$$
$$\frac{2}{2} = \frac{1}{2} + \frac{1}{2} = 2 \times \frac{1}{2},$$
$$\frac{3}{2} = \frac{1}{2} + \frac{1}{2} + \frac{1}{2} = 3 \times \frac{1}{2},$$
etc.

Where the number on top (the numerator) tells us how many copies we have of the unit fraction, and the number on the bottom (the denominator) indicates the size; halves, thirds, fourths, etc., of what we are counting.

We can now associate a numerical value with these multiples of parts of a whole.

Using our number line and starting with two to divide the distance between the whole numbers into halves we have:

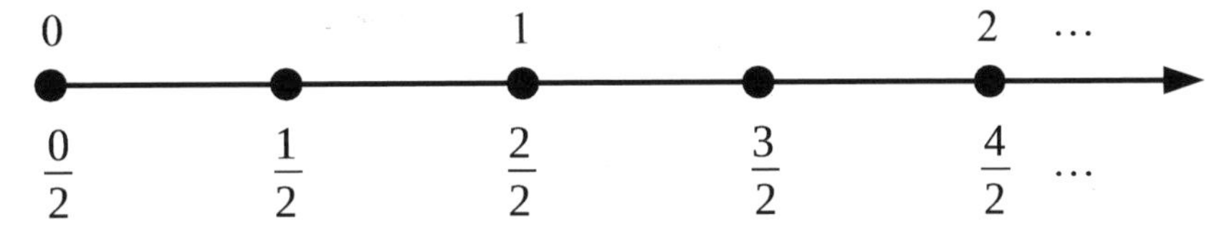

The first thing we notice is that we now have two equivalent ways of representing the same numerical value, or location, on the number line.

For example, we can either write 0, or we can write $\frac{0}{2}$. Meaning if we take zero and divide it in half we still have zero.

Next, we can either write 1 or we can write $\frac{2}{2}$. Meaning if we take two objects and divide that in half we have one object.

Continuing, we can either write 2 or we can write $\frac{4}{2}$. Again, two is equivalent to taking four objects and dividing that in half.

We don't have to stop here.

EXAMPLE 1:

Write equivalent fractions for 3, 4, 5, 6, using halves.

$$3 = \frac{6}{2}, \quad 4 = \frac{8}{2}, \quad 5 = \frac{10}{2}, \quad \text{and} \quad 6 = \frac{12}{2}$$

Another way to view an equivalent fraction is a different fraction that puts you at the same location on the number line.

The next thing we notice is that we can now think of the symbol for representing parts of a whole as a number that can identify a location on a number line.

$\frac{1}{2}$ represents the location halfway between 0 and 1 on the number line,

$\frac{3}{2}$ represents the location halfway between 1 and 2 on the number line.

Continuing in this way, $\frac{5}{2}$ would represent the location halfway between 2 and 3 on the number line, etc.

EXAMPLE 2:

What are the fractions representing numbers halfway between 7 and 8, between 9 and 10?

$$\frac{15}{2} \quad \text{and} \quad \frac{19}{2} \text{ respectively.}$$

We also see that there is an order to these fractions, just like there is an order to the counting numbers. Fractions can be less than or greater than other fractions and their position can be precisely identified on a number line. They really are numbers in their own right.

Now let's consider thirds with a number line.

$$\begin{array}{ccccccccc} 0 & & & 1 & & & 2 & \dots \\ \bullet & \bullet & \bullet & \bullet & \bullet & \bullet & \bullet & \rightarrow \\ \frac{0}{3} & \frac{1}{3} & \frac{2}{3} & \frac{3}{3} & \frac{4}{3} & \frac{5}{3} & \frac{6}{3} & \dots \end{array}$$

One thing to notice is that we now have more equivalent ways to write 0, 1, and 2.

For example, we can either write 0, or we can write $\frac{0}{3}$

We can either write 1, or we can write $\frac{3}{3}$

We can either write 2, or we can write $\frac{6}{3}$

Since they refer to the same location on the number line, they have an equivalent numerical value.

Furthermore, we can say that $\frac{0}{2}$ is equivalent to $\frac{0}{3}$,

and that $\frac{2}{2}$ is equivalent to $\frac{3}{3}$

and $\frac{4}{2}$ is equivalent to $\frac{6}{3}$

Continuing in this way we establish a pattern of equivalent fractions for our counting numbers:

$$0 = \frac{0}{2} = \frac{0}{3} = \frac{0}{4} = \frac{0}{5}, \ \dots$$

$$1 = \frac{2}{2} = \frac{3}{3} = \frac{4}{4} = \frac{5}{5}, \ \dots$$

$$2 = \frac{4}{2} = \frac{6}{3} = \frac{8}{4} = \frac{10}{5}, \ \ldots$$

$$3 = \frac{6}{2} = \frac{9}{3} = \frac{12}{4} = \frac{15}{5}, \ \ldots$$

EXAMPLE 3:

Write equivalent fractions for 4, 5, 6, and 7 using thirds.

$$4 = \frac{12}{3}$$

$$5 = \frac{15}{3}$$

$$6 = \frac{18}{3}$$

$$7 = \frac{21}{3}$$

Each fraction above represents the exact same place on the number line as the other equal fractions and whole numbers.

Now let's consider fourths with a number line.

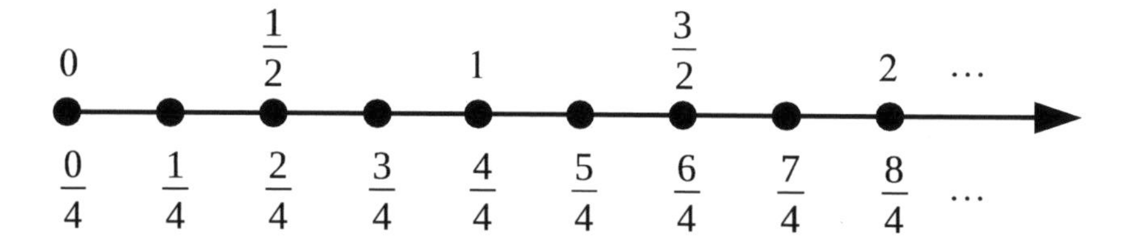

One thing we notice is that we now have more equivalent ways to write $\frac{1}{2}$ and $\frac{3}{2}$

For example, we can either write $\frac{1}{2}$ or we can write $\frac{2}{4}$

Similarly, we can either write $\frac{3}{2}$ or we can write $\frac{6}{4}$

Since they refer to the same location on the number line, they have an equivalent numerical value.

Continuing in this way we establish a pattern of equivalent fractions for $\frac{1}{2}, \frac{3}{2}, \frac{5}{2}, \frac{7}{2},$ etc.

$$\frac{1}{2} = \frac{2}{4} = \frac{3}{6} = \frac{4}{8}, \ \ldots$$

$$\frac{3}{2} = \frac{6}{4} = \frac{9}{6} = \frac{12}{8}, \ \ldots$$

$$\frac{5}{2} = \frac{10}{4} = \frac{15}{6} = \frac{20}{8}, \ \ldots$$

$$\frac{7}{2} = \frac{14}{4} = \frac{21}{6} = \frac{28}{8}, \ \ldots$$

EXAMPLE 4:

Write equivalent fractions for 4, 5, 6, and 7 using fourths.

$$4 = \frac{16}{4}$$

$$5 = \frac{20}{4}$$

$$6 = \frac{24}{4}$$

$$7 = \frac{28}{4}$$

Again, each fraction above represents the exact same place on the number line as the other equivalent fractions.

Using Multiplication and Division to Find Equivalent Fractions

To learn more about equivalent fractions we use the fraction bar (tape diagram), along with our number line. In the example below, we start by taking a whole and breaking it into three equal parts. We then broke two of the three parts each into four equal parts. From the picture below we can see that

$\frac{2}{3}$ is equivalent to $\frac{8}{12}$, or stated more directly $\frac{2}{3} = \frac{8}{12}$

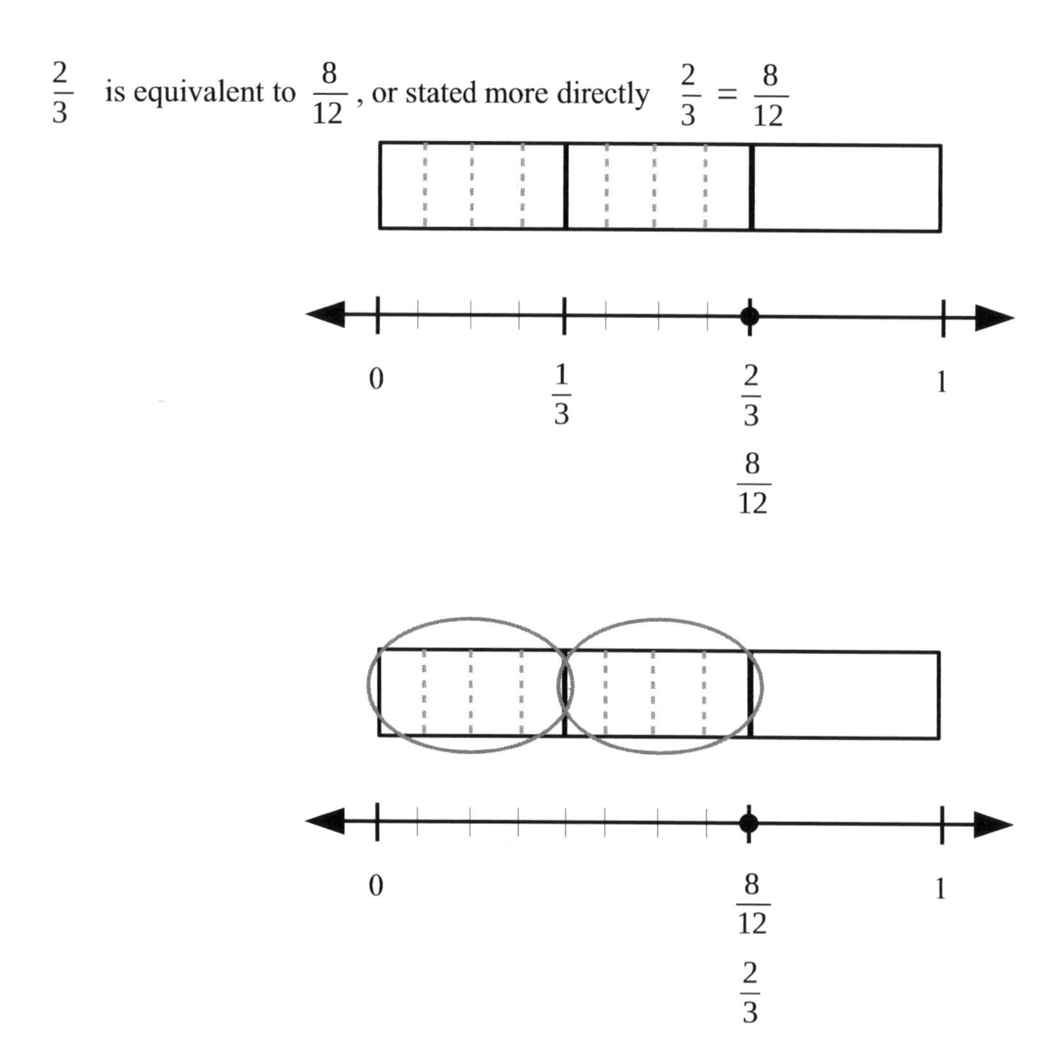

We can also show this using multiplication. We simply start with $\frac{2}{3}$ and multiply the top number 2 (called the numerator) by 4, and the bottom number 3 (called the denominator) also by 4, and get the equivalent fraction $\frac{8}{12}$.

$$\frac{2}{3} = \frac{2 \times 4}{3 \times 4} = \frac{8}{12}$$

In a similar way we can start with $\frac{8}{12}$, and take the numerator and divide by 4, and do the same to the denominator and get the equivalent fraction $\frac{2}{3}$.

$$\frac{8}{12} = \frac{8 \div 4}{12 \div 4} = \frac{2}{3}$$

The number line and the fraction bar illustrate the concept of what we are doing, while the multiplication and division approach is the "shorthand" algorithm we will use to find equivalent fractions more quickly.

We show this in the figures below.

$$\frac{2}{3} = \frac{2 \times 4}{3 \times 4} = \frac{8}{12}$$

0 $\quad \frac{1}{3} \quad \frac{2}{3} \quad$ 1

$\frac{8}{12}$

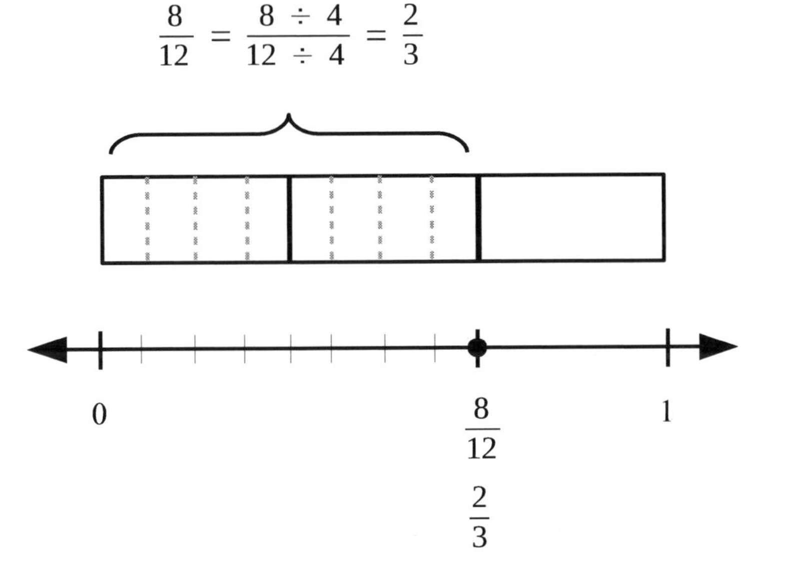

Area Model of Fractions

To help us understand this even better we develop an area model to show the relationship between multiplication and an equivalent fraction. For example, let's show $\frac{1}{2}$ using an area model. For this we simply draw a box and shade in half the box to represent our fraction, as shown below.

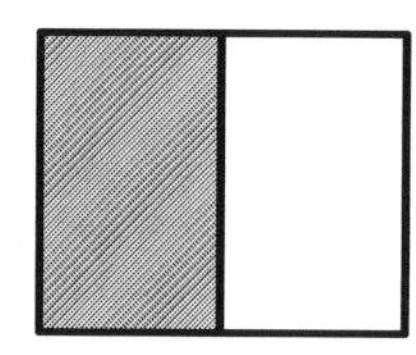

Now if we want to rewrite $\frac{1}{2}$ using four times as many units, we can simply divide up the one-half area into four equal sub-areas as shown.

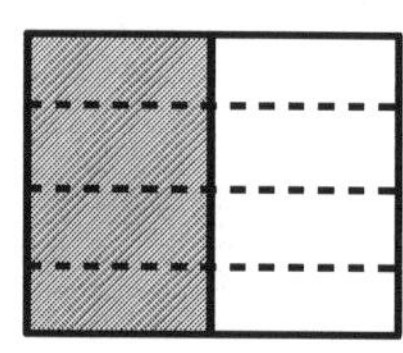

Then we can see the equivalence that $\frac{1}{2} = \frac{4}{8}$. Which we could also have obtained through multiplication: $\frac{1}{2} = \frac{1 \times 4}{2 \times 4} = \frac{4}{8}$

In the same way the following area chart,

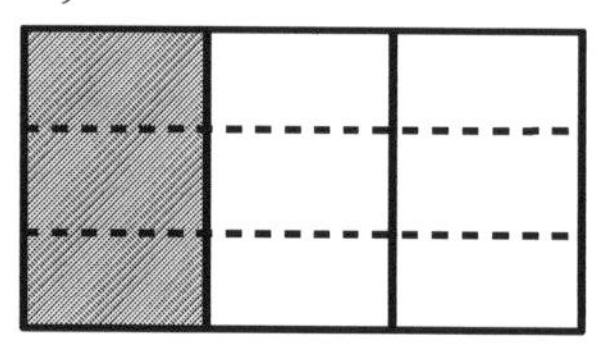

shows that

$$\frac{1}{3} = \frac{1 \times 3}{3 \times 3} = \frac{3}{9}$$

Another example:

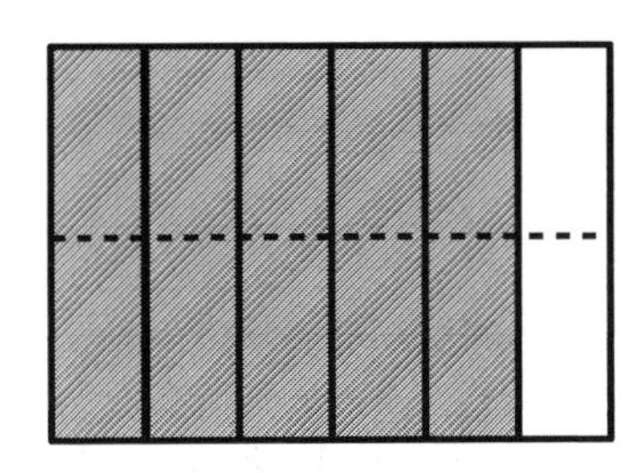

$$\frac{5}{6} = \frac{5 \times 2}{6 \times 2} = \frac{10}{12}$$

One more example:

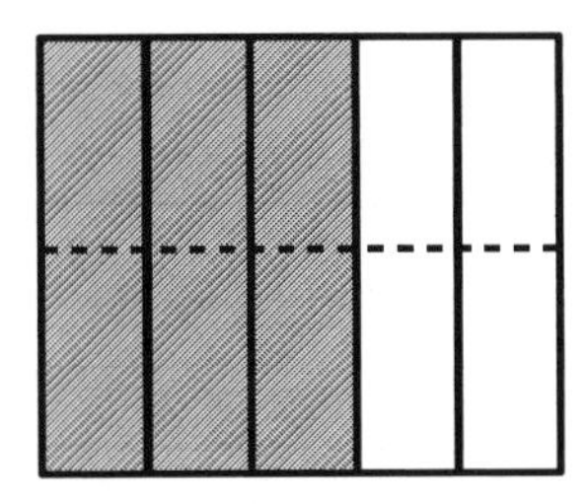

$$\frac{3}{5} = \frac{3 \times 2}{5 \times 2} = \frac{6}{10}$$

EXERCISES 3.4

Fractions as Numbers and Equivalent Fractions

1. Write equivalent fractions for 3, 4, 5, 6, using thirds.
2. Write equivalent fractions for 3, 4, 5, 6, using fourths.
3. Write equivalent fractions for 3, 4, 5, 6 using fifths.
4. Write equivalent fractions for 3, 4, 5, 6 using sixths.
5. What are the fractions representing numbers halfway between 6 and 7, between 11 and 12?
6. What are the fractions representing numbers halfway between 8 and 9, between 13 and 14?
7. Show using a tape diagram and number line that $\frac{2}{5}$ is equivalent to $\frac{4}{10}$. Then show through either multiplication or division that $\frac{2}{5} = \frac{4}{10}$
8. Show using a tape diagram and number line that $\frac{4}{5}$ is equivalent to $\frac{12}{15}$. Then show through either multiplication or division that $\frac{4}{5} = \frac{12}{15}$

9. Show using a tape diagram and number line that $\frac{9}{18}$ is equivalent to $\frac{1}{2}$. Then show through either multiplication or division that $\frac{9}{18} = \frac{1}{2}$

10. Show using a tape diagram and number line that $\frac{9}{15}$ is equivalent to $\frac{3}{5}$. Then show through either multiplication or division that $\frac{9}{15} = \frac{3}{5}$

11. Use an area chart to show that $\frac{2}{3} = \frac{6}{9}$

12. Use an area chart to show that $\frac{5}{6} = \frac{20}{24}$

13. Use an area chart to show that $\frac{4}{7} = \frac{8}{14}$

14. Use an area chart to show that $\frac{3}{4} = \frac{12}{16}$

15. Use the area chart to show that $\frac{2}{3} = \frac{12}{18}$

16. Use the area chart to show that $\frac{4}{5} = \frac{12}{15}$

3.5 COMPARING FRACTIONS–Order and Fractions

In this lesson we learn how to compare two fractions. That is, we learn how to determine if two fractions are equal or if one fraction is greater than or less than another fraction. A larger (greater than) fraction is a fraction that is to the right of the other fraction on the number line, while a smaller fraction (less than) is a fraction that is to the left of the number on the number line. We call this the order of fractions on the number line.

For example, we will learn how to determine that $\frac{2}{5}$ comes before $\frac{4}{7}$ on the number line as shown below. This is often not an easy task, so we need a good set of approaches to be able to determine this.

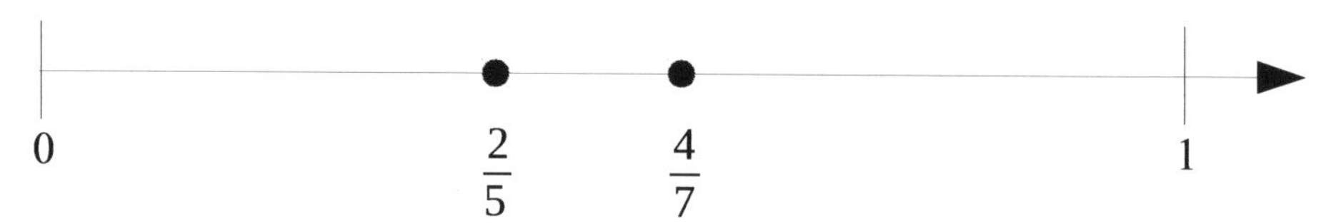

We will develop three different ways to compare two fractions. We will learn how to:

1. Compare our two fractions to a third more basic common fraction (such as $\frac{1}{2}$, $\frac{1}{3}$, or $\frac{1}{4}$) to determine the order of the two fractions.

2. Write the two fractions as equivalent fractions with an equivalent numerator (top) term to determine the order of the fractions.

3. Write the two fractions as two equivalent fractions with the same denominator (bottom) term to determine the order of the fractions.

These are 3 different techniques that can be used. It is up to us to pick which one is easiest to implement, for the problem at hand. Often times that is the hardest part; picking a technique.

Using Common Fractions to Compare two Fractions

To demonstrate this approach, we'll use the example at the beginning of this section.

EXAMPLE 1: Compare $\frac{2}{5}$ and $\frac{4}{7}$ determine their order. Which number comes before the other on a number line?

If we construct a tape diagram we can see by comparison below that $\frac{2}{5} < \frac{1}{2}$

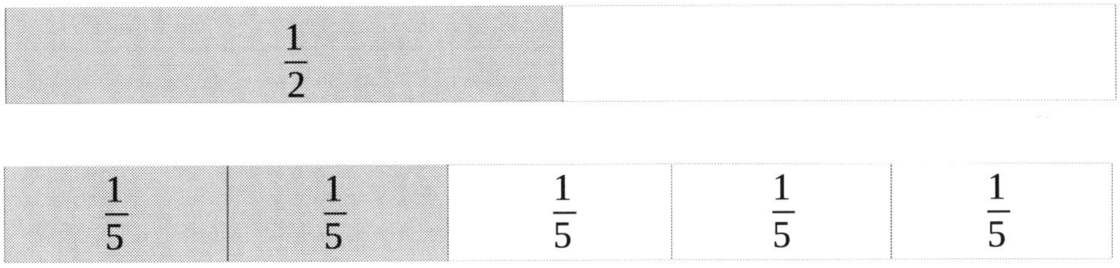

$\frac{1}{2}$	

$\frac{1}{5}$	$\frac{1}{5}$	$\frac{1}{5}$	$\frac{1}{5}$	$\frac{1}{5}$

In a similar way, we see that $\frac{1}{2} < \frac{4}{7}$

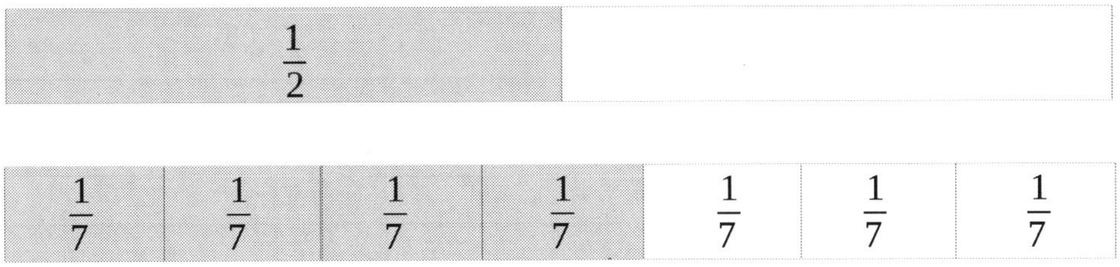

$\frac{1}{2}$	

$\frac{1}{7}$	$\frac{1}{7}$	$\frac{1}{7}$	$\frac{1}{7}$	$\frac{1}{7}$	$\frac{1}{7}$	$\frac{1}{7}$

Thus, since $\frac{2}{5} < \frac{1}{2}$ and $\frac{1}{2} < \frac{4}{7}$ we can say that $\frac{2}{5} < \frac{1}{2} < \frac{4}{7}$ or

$$\frac{2}{5} < \frac{4}{7}$$

EXAMPLE 2: Compare $\frac{4}{9}$ and $\frac{2}{7}$ to determine their order. Which number comes before the other on a number line?

Sketching a tape diagram, we see that $\frac{2}{7} < \frac{1}{3}$

$\frac{1}{3}$	$\frac{1}{3}$	$\frac{1}{3}$

$\frac{1}{7}$	$\frac{1}{7}$	$\frac{1}{7}$	$\frac{1}{7}$	$\frac{1}{7}$	$\frac{1}{7}$	$\frac{1}{7}$

In a similar fashion we see that $\frac{1}{3} < \frac{4}{9}$

$\frac{1}{3}$	$\frac{1}{3}$	$\frac{1}{3}$

$\frac{1}{9}$	$\frac{1}{9}$	$\frac{1}{9}$	$\frac{1}{9}$	$\frac{1}{9}$	$\frac{1}{9}$	$\frac{1}{9}$	$\frac{1}{9}$	$\frac{1}{9}$

Thus, since $\frac{2}{7} < \frac{1}{3}$ and $\frac{1}{3} < \frac{4}{9}$ we can say that $\frac{2}{7} < \frac{1}{3} < \frac{4}{9}$ or

$$\frac{2}{7} < \frac{4}{9}$$

Sometimes, it can be hard to find a basic fraction to compare two fractions with. This is why we need additional comparison techniques.

Using Equal Numerators to Compare two Fractions

The next technique is useful if both the top terms, the numerators, is the same for both of our fractions we are comparing. If it is then we use this technique. If this is the case then, the fraction with the smallest denominator is the larger fraction. We can demonstrate this with an example.

EXAMPLE 1: Compare the two fractions, $\frac{5}{7}$ and $\frac{5}{12}$. First, we note that they both have the same number, 5, in the numerator (top).

If we set up two tape diagrams, as shown below, we can see that $\frac{5}{7} > \frac{5}{12}$

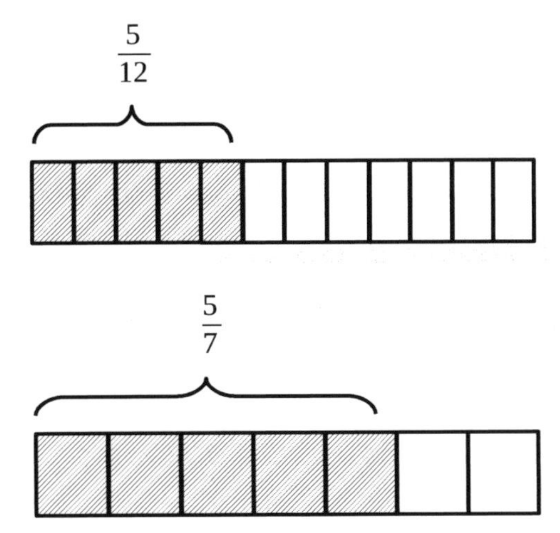

EXAMPLE 2: Compare the two fractions $\frac{9}{10}$ and $\frac{9}{12}$

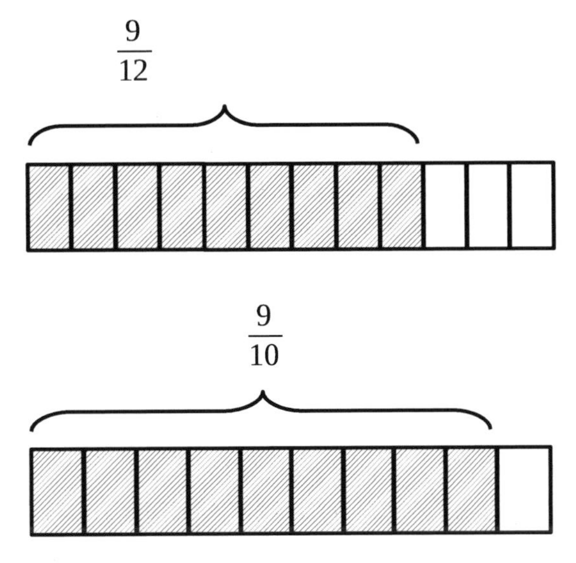

From this we can see that $\frac{9}{10} > \frac{9}{12}$

We can also use this method if the numerators can be made equivalent by a simple multiplication.

In this method for finding the order of two fractions, we use the multiplication approach developed earlier in this handout. We discovered previously that we get equivalent fractions if we multiply both the numerator and denominator by the same factor.

EXAMPLE 3:

To compare $\frac{2}{5}$ and $\frac{4}{7}$ we simply rewrite an equivalent fraction for one (or both) of the fractions that will give us the same numerator value for both fractions. Since 4 is a multiple of 2, we can multiply the fraction $\frac{2}{5}$ by 2, top and bottom, to obtain an equivalent fraction, and then compare.

Rewrite $\frac{2}{5}$ by multiplying the numerator and denominator by 2 to obtain,

$$\frac{2}{5} = \frac{2\times2}{2\times5} = \frac{4}{10}$$

This means we are actually comparing $\frac{4}{10}$ with $\frac{4}{7}$. We can see that $\frac{4}{7}$ is larger than $\frac{4}{10}$ since $\frac{1}{7}$ is larger than $\frac{1}{10}$ because we are dividing by a smaller number.

In terms of a tape diagram we can see this too.

$\frac{1}{10}$	$\frac{1}{10}$	$\frac{1}{10}$	$\frac{1}{10}$	$\frac{1}{10}$	$\frac{1}{10}$	$\frac{1}{10}$	$\frac{1}{10}$	$\frac{1}{10}$	$\frac{1}{10}$

$\frac{1}{7}$	$\frac{1}{7}$	$\frac{1}{7}$	$\frac{1}{7}$	$\frac{1}{7}$	$\frac{1}{7}$	$\frac{1}{7}$

Thus, we can conclude that

$$\frac{2}{5} < \frac{4}{7}$$

In summary, if we have equal numerators in our fractions, then the one with the larger denominator is the smaller (falls to the left of) of the two fractions. This hold true for equivalent fractions.

Using Equal Denominators to Compare two Fractions

Another approach focuses on the denominators of the two fractions. If the denominators are the same, or can be made to be the same by either a multiplication or division, we can easily determine which fraction is larger and which is smaller. Consider the following example.

EXAMPLE 1: Compare the two fractions $\frac{5}{8}$ and $\frac{7}{8}$

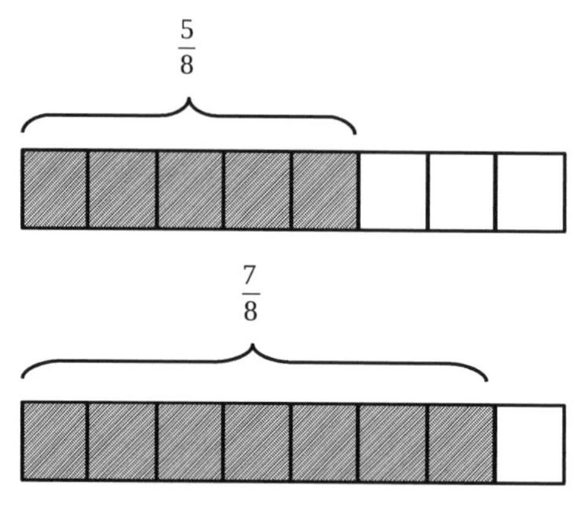

We can easily see that $\frac{5}{8} < \frac{7}{8}$

EXAMPLE 2: Compare the two fractions $\frac{9}{16}$ and $\frac{5}{8}$

In this example we see that 16 is a multiple of 8, so we multiply $\frac{5}{8}$ by 2 (top and bottom) and find the equivalent fraction, and compare the two.

$$\frac{5}{8} = \frac{2\times5}{2\times8} = \frac{10}{16}$$

Putting them on tape diagrams we see:

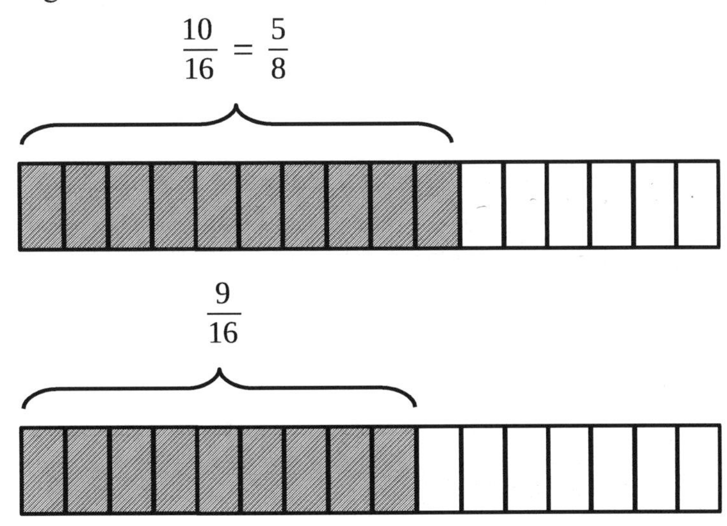

and conclude that $\frac{5}{8} > \frac{9}{16}$

EXERCISES 3.5

Comparing Fractions

Use a common fraction to determine which fraction is smaller.

1. $\frac{3}{4}$ and $\frac{2}{5}$
2. $\frac{2}{5}$ and $\frac{3}{7}$
3. $\frac{1}{5}$ and $\frac{2}{9}$
4. $\frac{2}{15}$ and $\frac{2}{11}$
5. $\frac{4}{5}$ and 1
6. $\frac{8}{5}$ and 1

Pick a technique to determine which fraction is smaller.

7. $\frac{7}{13}$ and $\frac{7}{9}$
8. $\frac{8}{9}$ and $\frac{8}{11}$
9. $\frac{3}{4}$ and $\frac{9}{13}$
10. $\frac{5}{9}$ and $\frac{20}{37}$
11. $\frac{7}{9}$ and $\frac{8}{9}$
12. $\frac{5}{11}$ and $\frac{7}{11}$
13. $\frac{4}{5}$ and $\frac{11}{15}$
14. $\frac{12}{18}$ and $\frac{5}{9}$
15. $\frac{8}{15}$ and 1
16. $\frac{14}{17}$ and 2
17. $\frac{2}{5}$ and $\frac{3}{8}$
18. $\frac{1}{8}$ and $\frac{3}{16}$

3.6 IMPROPER FRACTIONS AND MIXED NUMBERS

Up this this point we have only considered fractions to be made up of unit fraction parts of the segment of the number line from zero to one. However, fractions can be used to identify any point on the number line. This means we have to consider the possibility that we have more copies of the unit fraction than the size of the unit.

For example, what if we had the fraction $\frac{5}{4}$ which as we have learned previously is just 5 copies of the unit fraction $\frac{1}{4}$ placed end-to-end on the number line. We can see that after we place 4 of these $\frac{1}{4}$'s down we are at the location of 1 on the numbers line. Placing another $\frac{1}{4}$ next to this puts us to the left of 1 on the number line.

This is illustrated in the figures below using both the number line and the tape diagram.

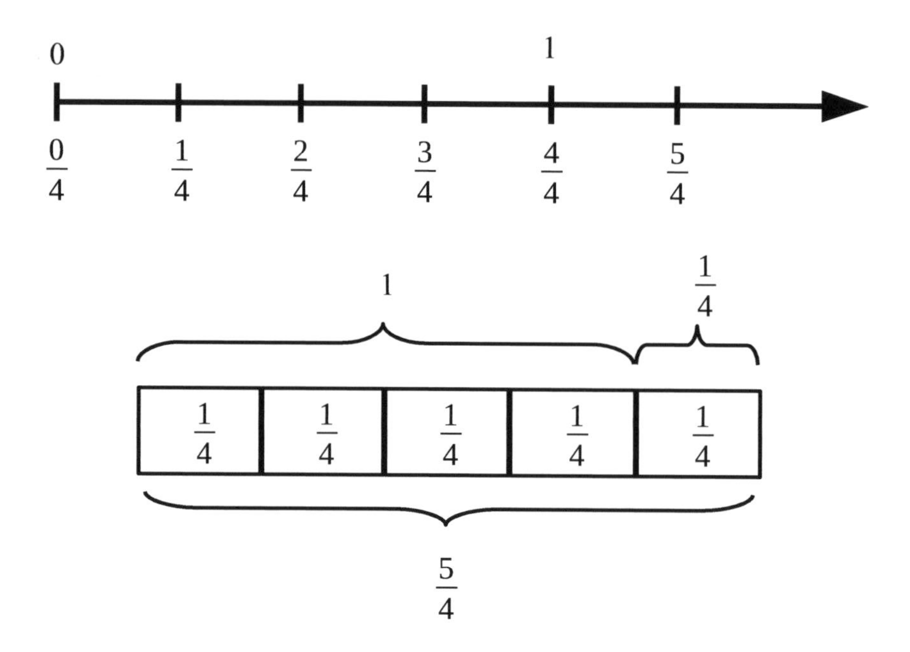

This is an example of something called an improper fraction. Whenever the number of copies is larger than the unit of the unit fraction, we call these fractions **improper fractions**. Fractions that are not improper fractions are called **proper fractions**.

Examples of improper fractions: $\frac{8}{3}, \frac{17}{11}, \frac{9}{2}$, etc. Improper fractions allow us to identify the location of numbers to the right of the number 1 on the number line.

Often times we would like a different way to think of these numbers, so that it makes it easier to get a sense of their "size," or location. This is accomplished using something called the **mixed number** form of a fraction.

The fraction $\frac{5}{4}$ as we have just shown, can be viewed as the whole number 1 and an additional $\frac{1}{4}$, or $\frac{5}{4} = 1$ and $\frac{1}{4}$ or just $1+\frac{1}{4}$.

Instead of writing the word "and", or the addition symbol, we simply drop this and write $1\frac{1}{4}$. This is what is called a mixed number. This number tells us that we are located at $\frac{1}{4}$ units beyond 1 on the number line.

Mixed numbers are useful when we want to get a sense for how big a number is, since they tell us how many whole number values we have to move along the number line, and then the additional unit fraction amount amount needed to move towards the next whole number. Improper fractions require us to do this extra step in our heads, so it is not as easy to get an immediate sense for the "size" of the number.

A **mixed number** is a number with two parts. The first part is a whole number part that tells us how many whole number unit values we move to the right on the number line, and a proper fraction part which tells us how many unit fractions we move towards the next whole number on the number line. The fraction part is never an improper fraction.

For example consider the fraction $\frac{8}{3}$. We know that this is equivalent to

$$\frac{8}{3} = \frac{3+3+2}{3} = \frac{3}{3} + \frac{3}{3} + \frac{2}{3} = 1 + 1 + \frac{2}{3} = 2 + \frac{2}{3},$$

since we are able to divide the number of copies of the unit fraction of $\frac{1}{3}$ into whole number values consisting of 3 copies of $\frac{1}{3}$. This gave us two whole units and an additional amount of $\frac{2}{3}$.

We can now simply rewrite this using the mixed number notation as:

$$2 + \frac{2}{3} = 2\frac{2}{3}$$

EXAMPLE 1: Write the following improper fractions as mixed numbers.

a) $\frac{27}{4}$ Solution: $\frac{27}{4} = \frac{26}{4}+\frac{3}{4} = 6+\frac{3}{4} = 6\frac{3}{4}$

a) $\frac{9}{5}$ Solution: $\frac{9}{5} = \frac{5}{5}+\frac{4}{5} = 1+\frac{4}{5} = 1\frac{4}{5}$

a) $\frac{37}{7}$ Solution: $\frac{37}{7} = \frac{35}{7}+\frac{2}{7} = 5+\frac{2}{7} = 5\frac{2}{7}$

We should also be able to translate a mixed number into an improper fraction.

EXAMPLE 2: Write the following mixed numbers as improper fractions.

a) $5\frac{4}{9}$ Solution: $5\frac{4}{9} = 5+\frac{4}{9} = \frac{9\times5}{9}+\frac{4}{9} = \frac{45}{9}+\frac{4}{9} = \frac{45+4}{9} = \frac{49}{9}$

a) $3\frac{2}{3}$ Solution: $3\frac{2}{3} = 3+\frac{2}{3} = \frac{3\times3}{3}+\frac{2}{3} = \frac{9}{3}+\frac{2}{3} = \frac{9+2}{3} = \frac{11}{3}$

a) $4\frac{7}{8}$ Solution: $4\frac{7}{8} = 4+\frac{7}{8} = \frac{8\times4}{8}+\frac{7}{8} = \frac{32}{8}+\frac{7}{8} = \frac{32+7}{8} = \frac{39}{8}$

EXERCISES 3.6

Write the improper fractions as mixed numbers.

1. $\frac{12}{7}$
2. $\frac{8}{5}$
3. $\frac{21}{8}$
4. $\frac{49}{20}$
5. $\frac{16}{11}$
6. $\frac{29}{5}$
7. $\frac{83}{10}$
8. $\frac{132}{25}$

Write the mixed numbers as improper fractions.

1. $3\frac{2}{7}$
2. $9\frac{1}{3}$
3. $4\frac{6}{7}$
4. $11\frac{2}{3}$
5. $1\frac{5}{11}$
6. $17\frac{3}{5}$
7. $28\frac{3}{4}$
8. $11\frac{8}{9}$

Chapter 3 Practice Test

Represent the unit fraction using a tape diagram.

1. $\frac{1}{9}$
2. $\frac{1}{6}$

Represent the fraction using a tape diagram and a number bond. Then decompose the fraction into unit fractions, and then in at least 2 different ways. Write a suitable addition and multiplication statement for each.

3. $\frac{5}{6}$
4. $\frac{4}{7}$
5. $\frac{9}{5}$

Break the given fractions into the given number of equal size pieces. Show a tape diagram. Then write addition and multiplication sentences that describe the relationship.

6. $\frac{1}{4}$ into 2 pieces
7. $\frac{1}{5}$ into 5 pieces
8. $\frac{4}{5}$ into 4 pieces
9. $\frac{4}{6}$ into 12 pieces

10. Write equivalent fractions for 3, 4, 5, 6 using fourths.

11. Write equivalent fractions for 3, 4, 5, 6 using fifths.

12. What are the fractions representing numbers halfway between 8 and 9, between 13 and 14?

13. Show using a tape diagram and number line that $\frac{9}{15}$ is equivalent to $\frac{3}{5}$. Then show through either multiplication or division that $\frac{9}{15}=\frac{3}{5}$.

14. Use an area chart to show that $\frac{4}{7}=\frac{8}{14}$

Use a common fraction to determine which fraction is smaller

15. $\frac{3}{7}$ and $\frac{3}{5}$

16. $\frac{8}{5}$ and $\frac{3}{11}$

Pick a technique to determine which fraction is smaller.

17. $\frac{4}{5}$ and $\frac{5}{11}$

18. $\frac{8}{9}$ and $\frac{8}{11}$

19. $\frac{8}{15}$ and 1

Write the improper fraction as mixed numbers.

20. $\frac{29}{6}$

21. $\frac{132}{25}$

Write the mixed numbers as improper fractions.

22. $11\frac{2}{5}$

23. $19\frac{3}{4}$

CHAPTER 4

Arithmetic of Fractions

4.1 ADDING FRACTIONS

Now that we see fractions as true numbers, that can be compared and calculated, we want to be able to perform arithmetic with them, just like we did with our whole numbers. In this section we introduce the process of how we add and subtract fractions. Ordinarily this can be quite confusing, especially since fractions look and behave quite different than our whole numbers. If, however, we keep the number line idea in mind, we can more easily explain and grasp the arithmetic of fractions.

Adding Fractions

Adding wholes is much easier, since we can simply count how many we have. Fractions, however, is like being given a bag of different length sticks and being required to determine how long it would be if they were laid end to end, but not be able to physically lay them all out and then measure them. The only way to be able to measure how long the entire bag is in this way, is to have to measure each piece with a common measuring device, and then add all these different lengths of measure together. In fractions, this common device is called the equivalent unit fraction. We have to transform each fraction into a sum of equivalent unit fractions, and then we can "count" how many equivalent unit fractions we have. Thus, we try and turn adding fractions into a process similar to adding whole numbers, but instead of counting wholes, we count unit fractions. Let's illustrate this with a few examples.

In our first example we start with a problem that is already set up for us, with each fraction based upon a common unit fraction. This means that they have the same denominator.

EXAMPLE 1: Add the following fractions: $\frac{3}{7} + \frac{2}{7}$

Since each fraction is based upon the common unit fraction of $\frac{1}{7}$, we can add all the seventh's together.

First let's show this visually using the number line.

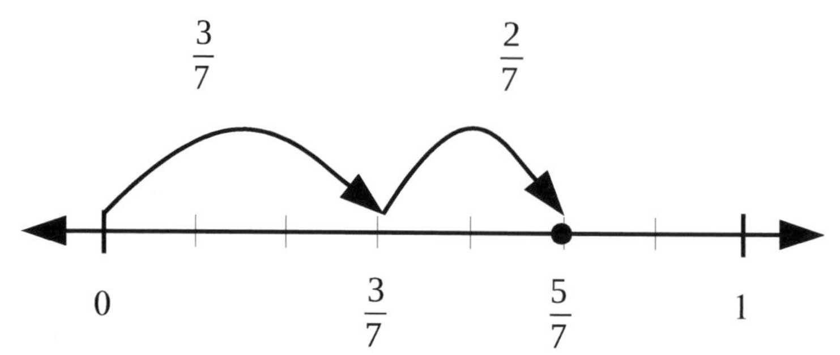

As you can see we start at zero then move $\frac{3}{7}$ unit to the right and then move another $\frac{2}{7}$ units to the right again, placing us at $\frac{5}{7}$

We can also see this using our tape diagrams.

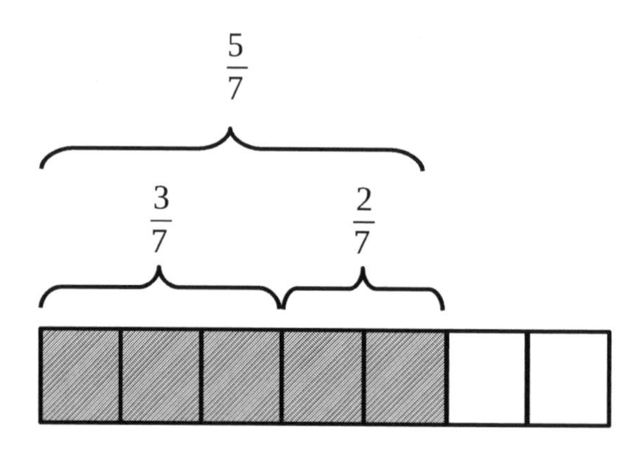

Finally, we show this with the unit fraction notation.

$$\frac{3}{7} = \frac{1}{7} + \frac{1}{7} + \frac{1}{7} \text{ and } \frac{2}{7} = \frac{1}{7} + \frac{1}{7}$$

so

$$\frac{3}{7} + \frac{2}{7} = \left(\frac{1}{7} + \frac{1}{7} + \frac{1}{7}\right) + \left(\frac{1}{7} + \frac{1}{7}\right) = \frac{5}{7}$$

EXAMPLE 2: Add the following fractions: $\frac{9}{16} + \frac{15}{16}$

Since each fraction is based upon the common unit fraction of $\frac{1}{16}$, we can add all the sixteenth's together.

Visually, using our number line.

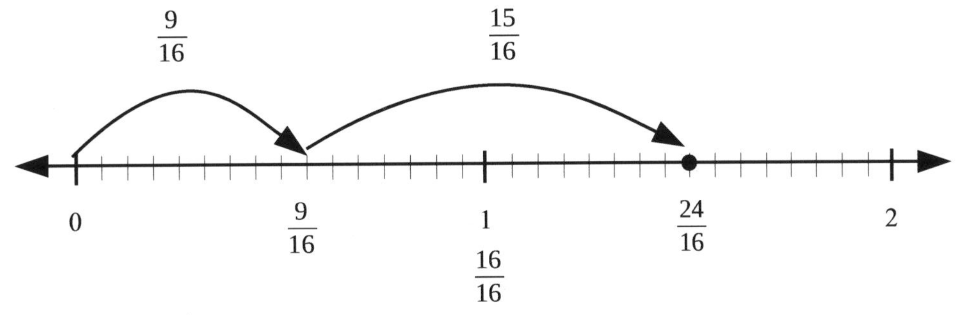

With the tape diagram we have

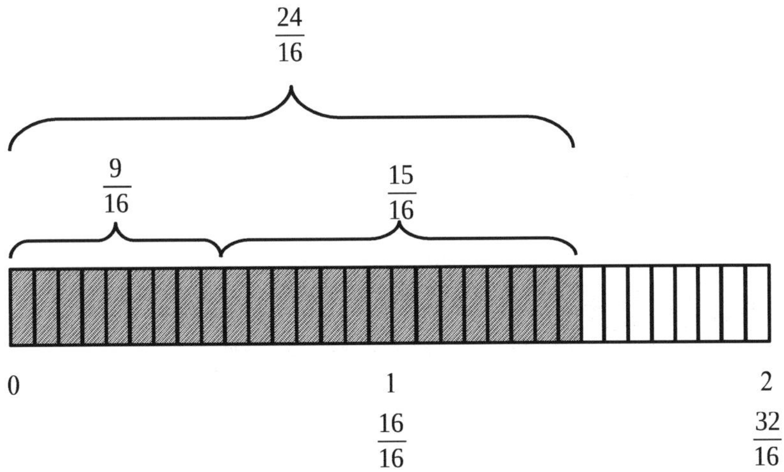

EXAMPLE 3: Add the following fractions: $\frac{3}{4} + \frac{5}{8}$

Since one fraction is based upon the unit fraction of $\frac{1}{4}$, and the other is based on the unit fraction of $\frac{1}{8}$ we have to find a common equivalent fraction before we can add them together.

Visually, using our number line.

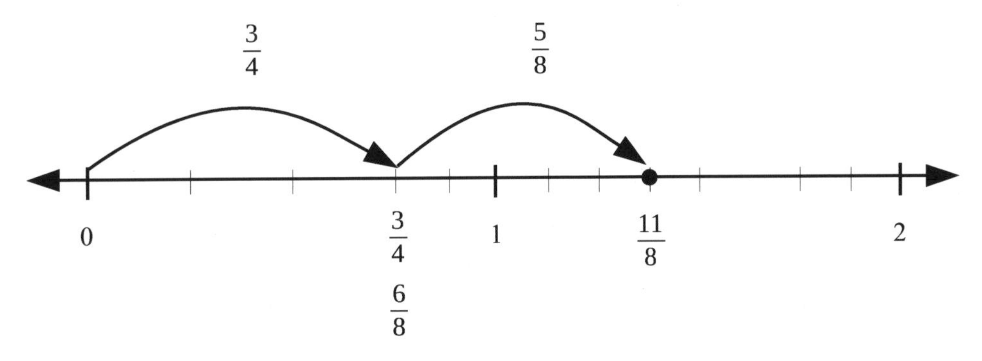

From this we see that $\frac{3}{4}$ is equivalent to $\frac{6}{8}$, so we can now add the fractions together as

$$\frac{6}{8} + \frac{5}{8} = \frac{11}{8}$$

With the tape diagram we have

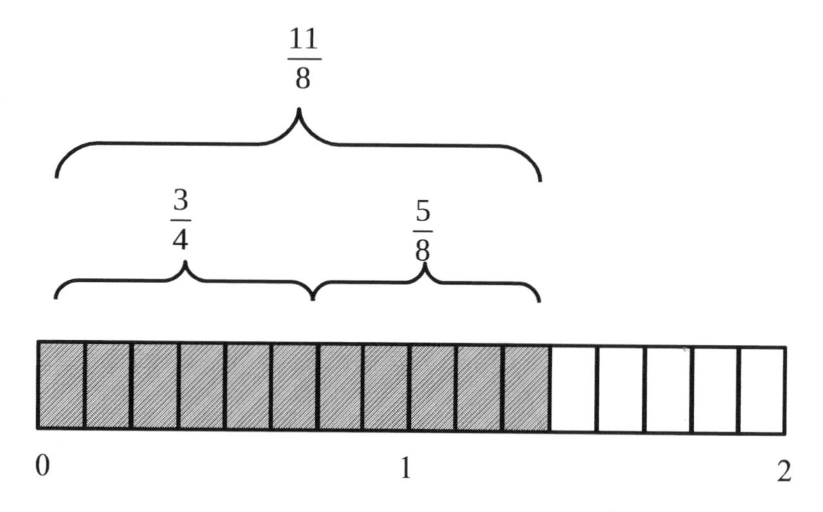

We can also use an area model to find the basic common unit fraction.

We first represent the two fractions.

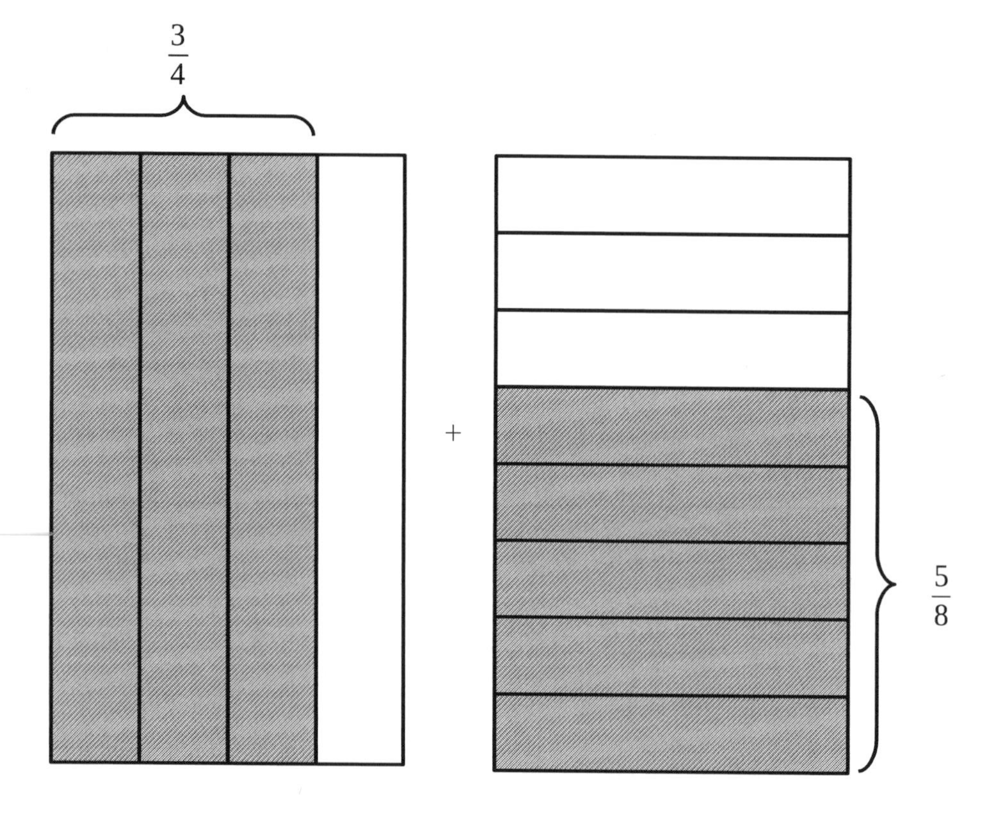

Next, we create a common unit fraction by dividing the ¾ in half, and add all the common 1/8 unit fractions together.

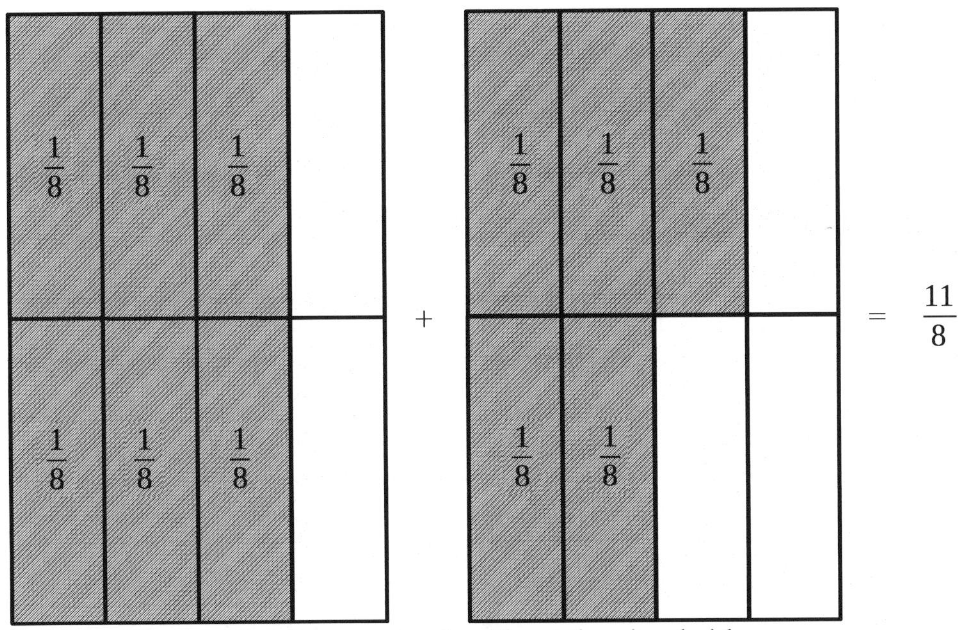

Finally we use our equivalent fraction approach without the visual aid.

In this problem we see that 4 and 8 are related, since $8 = 2 \times 4$. This means we only have to change one of the fractions into its equivalent form, and that is the $\frac{3}{4}$, which becomes,

$$\frac{3}{4} = \frac{3 \times 2}{4 \times 2} = \frac{6}{8}$$

Then we can write

$$\frac{3}{4} + \frac{5}{8} = \frac{6}{8} + \frac{5}{8} = \frac{11}{8}$$

We'll leave this answer as an improper fraction of $\frac{11}{8}$

EXAMPLE 4: Add the following fractions: $\frac{3}{7} + \frac{5}{6}$

In this problem we see that 7 and 6 are not related, meaning that they have no factors in common. Using our number line with two different scales (unit fractions) we can see the problem.

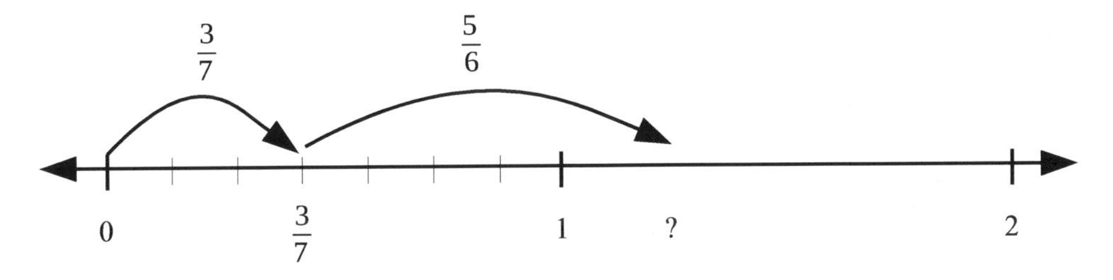

We are using two different unit fractions to add, so we don't know where to place our final point. Now we could work with the number line to find the common unit fraction, but we'll use the area model instead. (Note: the tape diagram is similar to the number line in its visualization of fractions.)

If we use our area model we can easily see the common unit fraction as shown below. First we visual represent our two fractions we are adding.

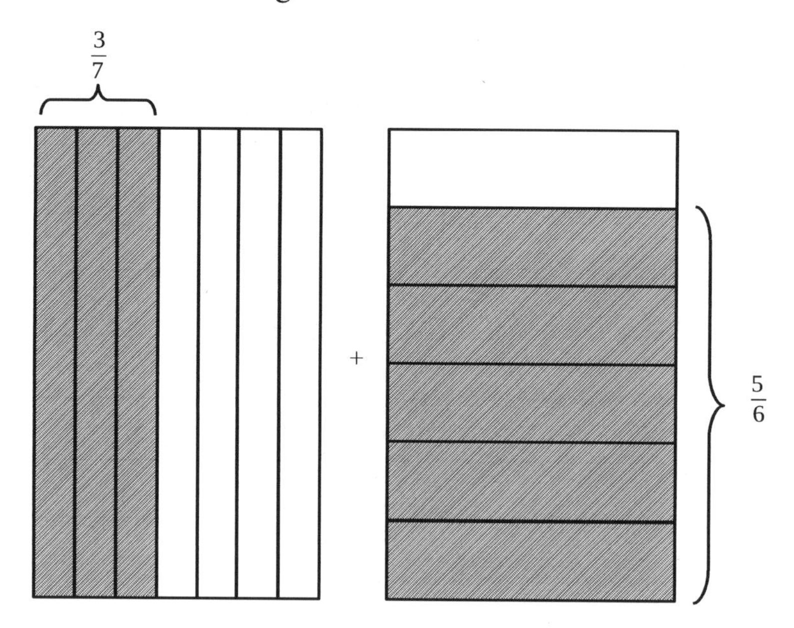

Next, if we divide the 3/7 into 6 equal parts as well as the 5/6 into 7 equal parts, we have the following.

$\frac{1}{42}$	$\frac{1}{42}$	$\frac{1}{42}$				
$\frac{1}{42}$	$\frac{1}{42}$	$\frac{1}{42}$				
$\frac{1}{42}$	$\frac{1}{42}$	$\frac{1}{42}$				
$\frac{1}{42}$	$\frac{1}{42}$	$\frac{1}{42}$				
$\frac{1}{42}$	$\frac{1}{42}$	$\frac{1}{42}$				
$\frac{1}{42}$	$\frac{1}{42}$	$\frac{1}{42}$				

$+$

$\frac{1}{42}$	$\frac{1}{42}$	$\frac{1}{42}$	$\frac{1}{42}$	$\frac{1}{42}$	$\frac{1}{42}$	$\frac{1}{42}$
$\frac{1}{42}$	$\frac{1}{42}$	$\frac{1}{42}$	$\frac{1}{42}$	$\frac{1}{42}$	$\frac{1}{42}$	$\frac{1}{42}$
$\frac{1}{42}$	$\frac{1}{42}$	$\frac{1}{42}$	$\frac{1}{42}$	$\frac{1}{42}$	$\frac{1}{42}$	$\frac{1}{42}$
$\frac{1}{42}$	$\frac{1}{42}$	$\frac{1}{42}$	$\frac{1}{42}$	$\frac{1}{42}$	$\frac{1}{42}$	$\frac{1}{42}$
$\frac{1}{42}$	$\frac{1}{42}$	$\frac{1}{42}$	$\frac{1}{42}$	$\frac{1}{42}$	$\frac{1}{42}$	$\frac{1}{42}$

$= \frac{53}{42}$

Now we can visual see why this approach works, but it would be extremely cumbersome if we had to do this every time. Instead we rely on the arithmetic approach, using the fraction notation and writing equivalent fractions numerically, to do our calculations. We only use the visual approach as a means for understanding the concept of adding fractions with unlike denominators.

We finish this problem using the arithmetic approach.

We still need to find the common equivalent unit fraction we need so that we can find out how many equivalent unit fractions they have between the both of them.

> If the denominators do not have any factors in common, we simply multiply the denominators together to get the smallest common equivalent denominator.

In this case it is

$$7 \times 6 = 42$$

This means we have to change both of the fractions into their equivalent forms with 42 in the denominator.

$$\frac{3}{7} \text{ becomes } \frac{3}{7} = \frac{3 \times 6}{7 \times 6} = \frac{18}{42}$$

Notice, for this fraction we simply multiply by the opposite factor of 6.

$$\frac{5}{6} \text{ becomes } \frac{5}{6} = \frac{5 \times 7}{6 \times 7} = \frac{35}{42}$$

In this fraction we multiplied by the opposite factor of 7 to get our equivalent fraction.

Then we can write

$$\frac{3}{7} + \frac{5}{6} = \frac{18}{42} + \frac{35}{42} = \frac{53}{42}$$

We'll leave this answer as an improper fraction of $\frac{53}{42}$

EXAMPLE 5: Add the following fractions: $\frac{7}{8} + \frac{11}{12}$

In this problem we will only demonstrate the arithmetic approach.

We first observe that 8 and 12 have a common factor of 4, since

$$8 = 2 \times 4$$

and

$$12 = 3 \times 4$$

We see that the uncommon factors are 2 and 3. The least common equivalent denominator is obtained by multiplying by the uncommon factors (2 and 3) and the common factor (4), or

$$2 \times 3 \times 4 = 24$$

> If the denominators do have factors in common, we simply multiply the uncommon factors by the common factors to get our common denominator.

This means we now have to change both of the fractions into their equivalent forms with 24 in the denominator.

$$\frac{7}{8} \text{ becomes } \frac{7}{8} = \frac{7 \times 3}{8 \times 3} = \frac{21}{24}$$

Notice, for this fraction we simply multiply by the opposite uncommon factor of 3.

$$\frac{11}{12} \text{ becomes } \frac{11}{12} = \frac{11 \times 2}{12 \times 2} = \frac{22}{24}$$

In this fraction we multiplied by the opposite uncommon factor of 2 to get our equivalent fraction.

Then we can write

$$\frac{7}{8} + \frac{11}{12} = \frac{21}{24} + \frac{22}{24} = \frac{43}{24}$$

We'll leave this answer as an improper fraction of $\frac{43}{24}$

EXERCISES 4.1

Adding Fractions

Add the following fractions with the standard algorithm and visually show the process using a number line and tape diagram.

1. $\frac{3}{8} + \frac{5}{8}$
2. $\frac{5}{12} + \frac{1}{12}$
3. $\frac{4}{9} + \frac{2}{9}$
4. $\frac{4}{11} + \frac{3}{11}$
5. $\frac{3}{10} + \frac{2}{5}$
6. $\frac{2}{7} + \frac{5}{14}$
7. $\frac{2}{9} + \frac{2}{3}$
8. $\frac{1}{4} + \frac{5}{12}$

Add the following fractions with the standard algorithm and visually show the process using an area model.

9. $\frac{2}{5} + \frac{3}{7}$
10. $\frac{2}{9} + \frac{4}{5}$
11. $\frac{2}{3} + \frac{2}{7}$
12. $\frac{1}{6} + \frac{3}{11}$

Add the following fractions with the standard algorithm

13. $\frac{3}{13} + \frac{5}{26}$
14. $\frac{5}{21} + \frac{1}{7}$
15. $\frac{7}{18} + \frac{8}{9}$
16. $\frac{3}{5} + \frac{4}{15}$
17. $\frac{9}{16} + \frac{7}{24}$
18. $\frac{5}{8} + \frac{7}{18}$
19. $\frac{3}{16} + \frac{5}{12}$
20. $\frac{9}{22} + \frac{8}{33}$

Addition of fractions applications

21. A farmer stated that he and his crew harvested $\frac{5}{8}$ of his strawberry field on day 1. They

harvested $\frac{1}{5}$ of the field on day 2. How much of the whole field did the farmer and his crew harvested on the two days?

22. Jonathan read $\frac{1}{2}$ of a book on Monday and then $\frac{1}{3}$ of the book on Tuesday. How much of the book has he read?
23. Juanita read $\frac{2}{5}$ of a book on Monday and then $\frac{1}{4}$ of the book on Tuesday. How much of the book has she read?
24. You completed half of your homework last night and one-quarter the night before. How much of your homework have you completed?

4.2 SUBTRACTING FRACTIONS

We have spent a great deal of time showing how to do addition of fractions. This should now make the concept of subtraction easier. Subtraction of fractions is very similar to addition. Only now we remove common unit fractions instead of adding them.

Let's see this through a few examples.

EXAMPLE 1: Subtract the following fractions: $\frac{6}{7} - \frac{4}{7}$

In this case the denominators are the same, so we already know our unit fraction is 1/7. Then we can visually see what this is through our number line.

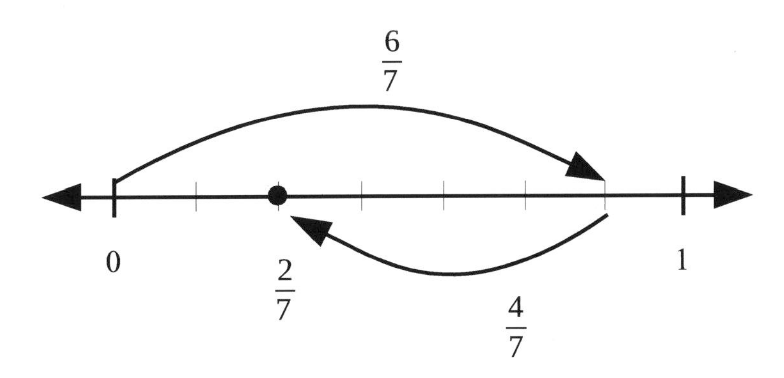

We first move 6/7 units to the right, then since we are subtracting we move 4/7 units to the left, leaving us at the location of 2/7.

This is similar to our tape diagram representation. We start with 6/7 and remove 4/7 of the tape.

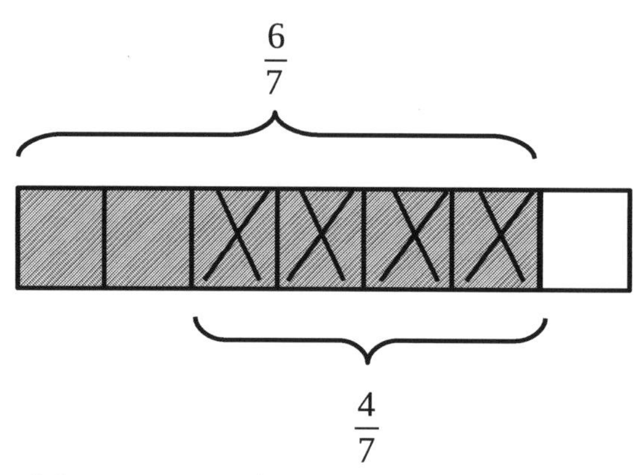

This leaves us with only 2/7 of the tape.

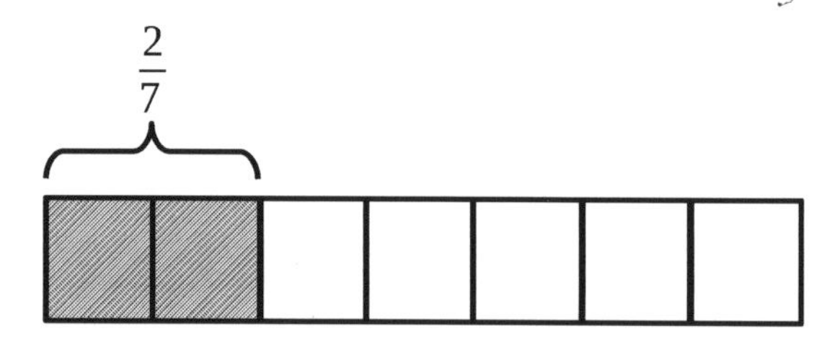

So our final answer is $\frac{2}{7}$

Arithmetically, to solve $\frac{6}{7} - \frac{4}{7}$ we notice that the denominators are the same, so we can simply subtract the numerators.

$$\frac{6}{7} - \frac{4}{7} = \frac{6-4}{7} = \frac{2}{7}$$

The above calculation is the standard algorithm, but the number line and the tape diagram explain why we can do this simple calculation.

EXAMPLE 2: Subtract the following fractions: $\frac{3}{4} - \frac{5}{8}$

We could show the same work visually using either the number line, tape diagram, or even the area model, but since we spent a lot of time doing this for addition, we will use the idea that it is the same, and with subtraction are removing, rather than adding, the unit fractions.

In this case the denominators are not the same, so we must obtain a common denominator (equivalent unit fraction) for both. We notice that we can simply change the 3/4's and rewrite the equivalent fraction of $\frac{6}{8}$.

Thus, we can obtain.

$$\frac{3}{4} - \frac{5}{8} = \frac{6}{8} - \frac{5}{8} = \frac{6-5}{8} = \frac{1}{8}$$

EXERCISES 4.2

Subtracting Fractions

Subtract the following fractions with the standard algorithm and visually show the process using a number line and tape diagram.

1. $\frac{5}{8} - \frac{1}{8}$
2. $\frac{5}{12} - \frac{1}{12}$
3. $\frac{4}{9} - \frac{2}{9}$
4. $\frac{4}{11} - \frac{3}{11}$
5. $\frac{2}{5} - \frac{3}{10}$
6. $\frac{4}{7} - \frac{5}{14}$
7. $\frac{7}{9} - \frac{2}{3}$
8. $\frac{3}{4} - \frac{5}{12}$

Subtract the following fractions with the standard algorithm and visually show the process using an area model.

9. $\frac{3}{5} - \frac{2}{7}$
10. $\frac{5}{9} - \frac{2}{5}$
11. $\frac{2}{3} - \frac{2}{7}$
12. $\frac{5}{6} - \frac{3}{11}$

Subtract the following fractions with the standard algorithm

13. $\frac{3}{13} - \frac{5}{26}$
14. $\frac{10}{21} - \frac{1}{7}$
15. $\frac{7}{18} - \frac{2}{9}$
16. $\frac{3}{5} - \frac{4}{15}$
17. $\frac{9}{16} - \frac{5}{24}$
18. $\frac{5}{8} - \frac{5}{18}$
19. $\frac{7}{16} - \frac{5}{12}$
20. $\frac{9}{22} - \frac{5}{33}$

Subtraction of fractions applications

21. Kaitlyn read $\frac{2}{5}$ of a book on Monday, then read $\frac{1}{4}$ of the same book on Tuesday, and the remainder of the book on Wednesday, What part of the book did she read on Wednesday?

22. Ethan ate $\frac{1}{4}$ of a huge cookie his mother made for him. Two hours later, he ate $\frac{1}{3}$ of the same cookie. What part of the cookie is left?

23. A student did $\frac{3}{5}$ of her homework on Saturday morning and $\frac{1}{4}$ of her homework in the afternoon. She said she would do the remaining homework in the evening. What part of her homework does she still have to do?

4.3 MULTIPLYING FRACTIONS

When we learned how to multiply whole numbers we thought of multiplication as repeated addition, and division as finding the number of times you were repeating the additions. Applying the operations of multiplication and division to fractions, we will use the same concepts. A number is a number, regardless of its form, so the operations should be the same conceptually. We start with multiplication.

Again, multiplication is the process of repeated additions. It is stated as

$$(\text{Number } 1)\times(\text{Number } 2)$$

and asks the question;

What number is obtained when (Number 2) is added to itself (Number 1) times?

Multiplying Whole Numbers Times Fractions

Let's start with the simplest form of multiplication, that of a whole number times a fraction. Now, we have already seen this in our multiplication statement for adding unit fractions. For example we saw that

$$9 \times \frac{1}{16} = \frac{9}{16}$$

We essentially multiplied the 9 times the 1 in the numerator of the unit fraction. This was because 9/16 was just 9 copies of 1/16. The same holds true if the term we are multiplying is not a unit fraction. Let's illustrate this with some examples.

EXAMPLE 1: Multiply the following: $4 \times \frac{3}{5}$

Visual Models

What this is telling us is that we have 4 copies of 3/5. We can show this, with either our number line,

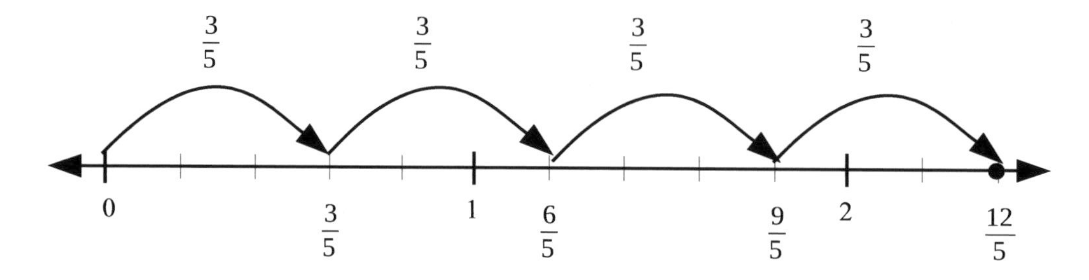

or our tape diagram

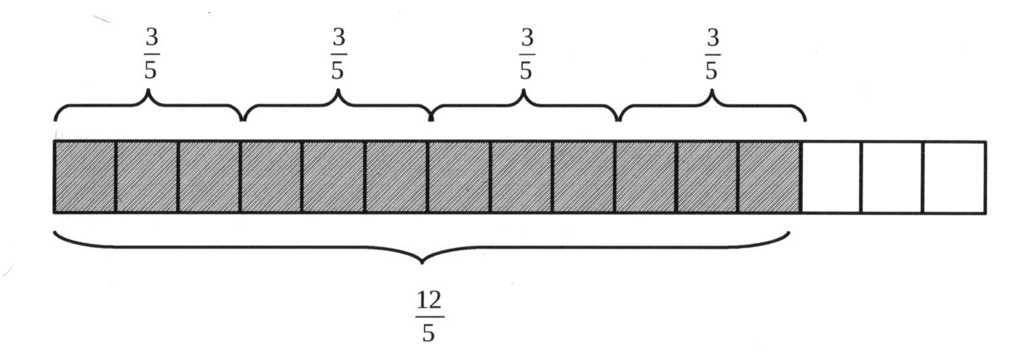

Standard Algorithm
Arithmetically we have the standard algorithm which says we just multiply the numerator by the whole number we are multiplying by.

$$4 \times \frac{3}{5} = \frac{4 \times 3}{5} = \frac{12}{5}$$

We have left our final answer as an improper fraction. We could change it to a mixed number and get $2\frac{2}{5}$.

We should point out that the operation of multiplication is what we call commutative. That means $3 \times 9 = 9 \times 3$. We can reverse the order in which we multiple and get the same answer. The reason why we mention this now is because if we see the following multiplication problem $\frac{5}{7} \times 6$ we can always rewrite it as $6 \times \frac{5}{7}$ and use the same interpretation and approach for multiplying we just introduced. Both expressions mean that we have 6 copies of $\frac{5}{7}$.

Furthermore, we should also point out that if we don't use the above analogy, there is another analogy we will use later on in interpreting $\frac{5}{7} \times 6$. This expression also means that we are trying to found out what number is $\frac{5}{7}$ of 6. Depending upon the type of problem we are working on, one analogy or the other will be easier, so we need to understand both. Thus, we explain it in detail in the following.

Multiplying Fractions Times Whole Numbers

EXAMPLE 2: As we just said multiplying a whole number by a fraction, such as $\frac{5}{7} \times 14$, is the same as asking the question; $\frac{5}{7}$ of 14 is equal to what number?

Visual Model
Let's consider the concept visually using our tape diagram. Here, we will take the number 14 and

divide it into 7 equal parts and then just count 5 of them for $\frac{5}{7}$ of 14.

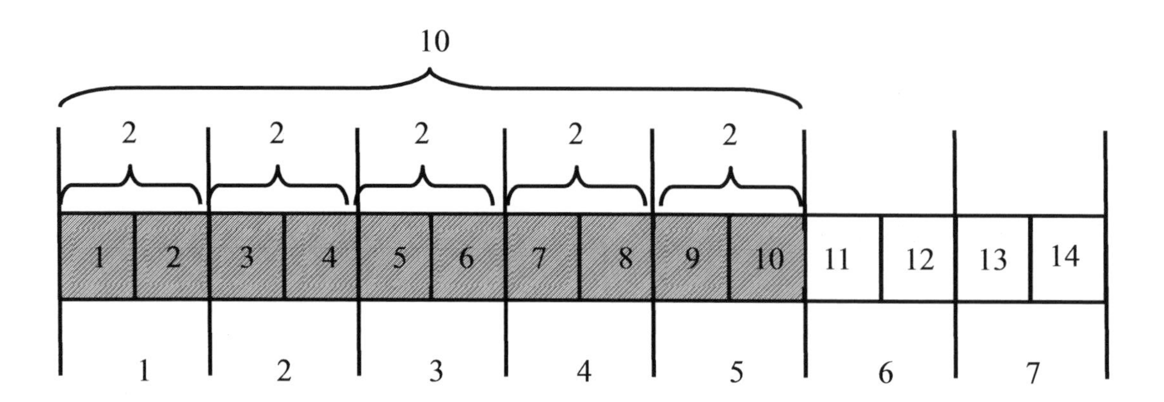

This shows us that $\frac{5}{7}$ of 14 is equal to 10.

Standard Algorithm

To compute using the standard algorithm we simply multiply the 5 times the 14 and divide by 7,

$$\frac{5}{7} \times 14 = \frac{5 \times 14}{7} = \frac{70}{7} = \frac{10 \times \cancel{7}}{\cancel{7}} = 10$$

Notice how we get the same result using our tape diagram.

EXAMPLE 2: $\frac{3}{5}$ of 35 is what number?

Visual Model

Using our tape diagram we interpret this as.

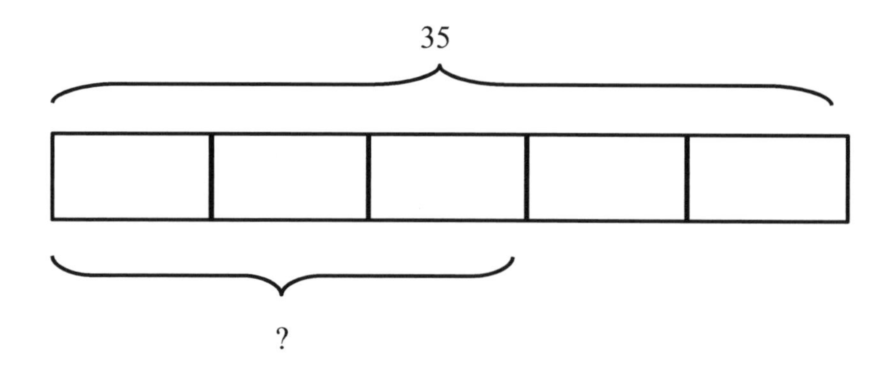

Thus, we must break 35 into 5 equal parts and count 3 of them.

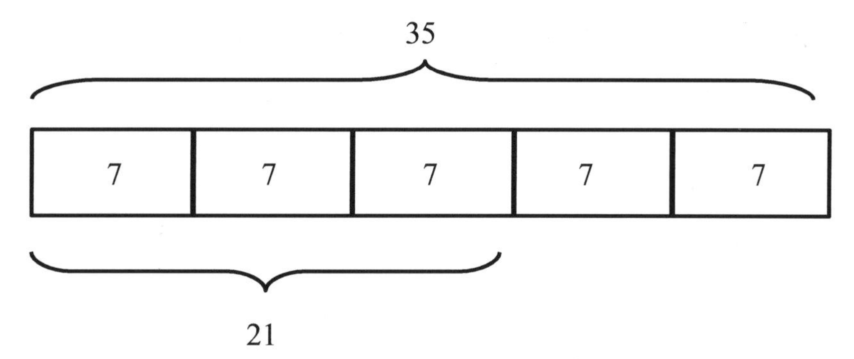

So we see our answer is 21.

Standard Algorithm
Again, if we use the standard algorithm we get.

$$\frac{3}{5} \times 35 = \frac{3 \times 35}{5} = \frac{115}{5} = \frac{21 \times 5}{5} = 21$$

In these examples the answer worked out to be a whole number. Most of the time the result is a fraction, though.

EXAMPLE 3: $\frac{2}{7}$ of 9 is what number?

Standard Algorithm
As we can see with this problem, our answer will not be a whole number. Let's start with the standard algorithm, and then try and justify it with another approach.

$$\frac{2}{7} \times 9 = \frac{2 \times 9}{7} = \frac{18}{7}$$

The final answer is the fraction $\frac{18}{7}$.

Visual Model
To justify this, consider the area model. We start with 9, and then break it into 7 equal parts, and then choose only two of the 7 parts (the shaded region below). Each smaller square is actually $\frac{1}{7}$. There are 18, $\frac{1}{7}$ squares shaded in for a total of $\frac{18}{7}$. Thus, the multiplication is just 18 repeated additions of the unit fraction $\frac{1}{7}$.

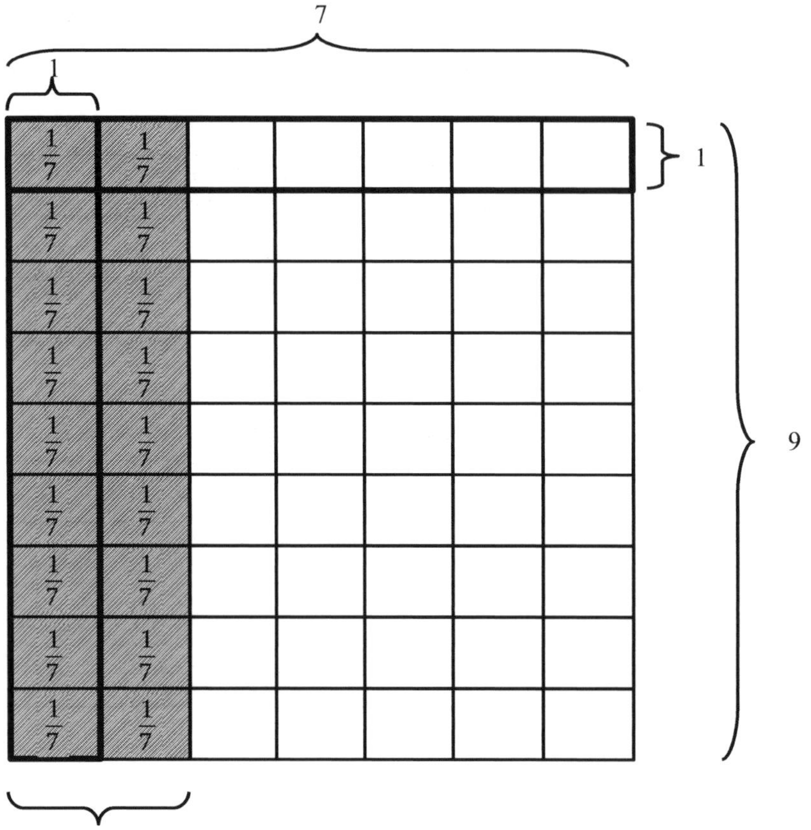

$$\frac{2}{7} \text{ of } 9 = 18 \times \frac{1}{7} = \frac{18}{7}$$

As we can see we get the same answer. This is our visual justification of the standard algorithm. This is why it works. For more complicated problems like this, however, we will use the standard algorithm.

Multiplying Fractions Times Fractions

We now consider multiplying a fraction by another fraction, such as $\frac{4}{5} \times \frac{2}{3}$. Again, this is still the multiplication operation, so the same ideas and concepts we've been discussing in this section still apply. Multiplication is still a repeated addition. What we are adding, and how many times, however, is not as obvious. In what follows we will attempt to explain it to you. Let's look at an example.

EXAMPLE 4: $\frac{4}{5}$ of $\frac{2}{3}$ is what number? Or $\frac{4}{5} \times \frac{2}{3}$ equals what number?

In this case we will start with an area interpretation before using the standard algorithm. We first want to understand the concept before we learn the short cut (standard algorithm) in finding our results. A conceptual understanding should always come first.

Visual Model

In considering this type of problem, $\frac{4}{5} \times \frac{2}{3}$, we will use the second interpretation of multiplying by a fraction in the beginning of this section. We have already shown that multiplying by a fraction is equivalent to saying we want to find the number that is $\frac{4}{5}$ of $\frac{2}{3}$. Recall that we have already shown that this is just another way to think of multiplication as repeated addition, but addition of specific unit fractions that we have to find. For this problem we shall start with $\frac{2}{3}$

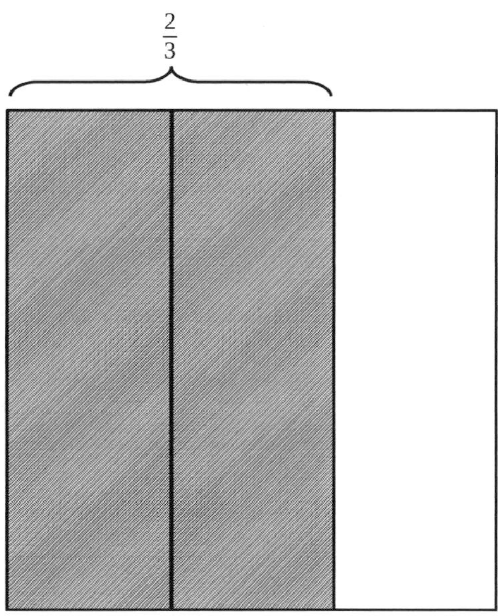

We then break the $\frac{2}{3}$ into fifths, and count only $\frac{4}{5}$ of the $\frac{2}{3}$ as shown below.

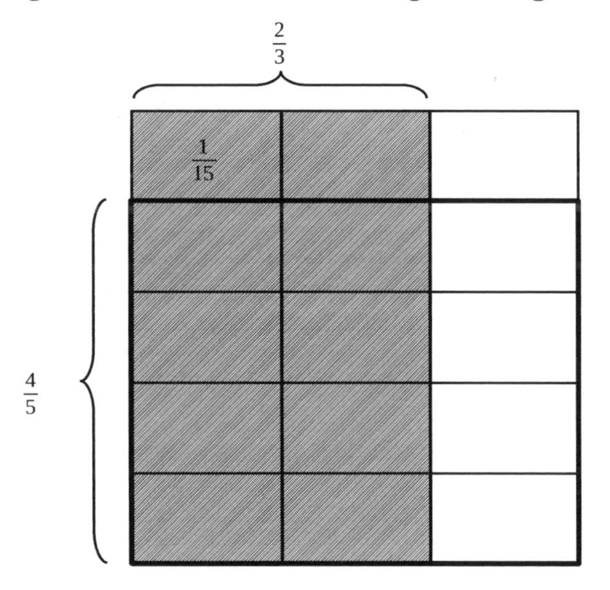

When we break the $\frac{2}{3}$ into fifths we see that the unit fraction we are adding is actually $\frac{1}{15}$.

If we count the overlap of the shaded region with the $\frac{4}{5}$ box we count 8 boxes. Thus, we are actually adding $\frac{1}{15}$ 8 times, for a final answer of $\frac{8}{15}$. Again, we can see that multiplication is still a repeated addition, even when the multipliers are fractions.

Standard Algorithm

Let's now see how this works for the standard algorithm.

$$\frac{4}{5} \times \frac{2}{3} = \frac{4 \times 2}{5 \times 3} = \frac{8}{15}$$

Again, the same final answer. You can see why standard algorithms were developed, as they make getting the final answer quite simple. The problem is when people use the standard algorithm and think this tells them what they are multiplying and what are the repeated unit fractions they are working with. It really doesn't. This is why so many students mess up the multiplication of fractions. They have learned so many standard algorithms that they all begin to merge together in their minds, so sometimes they mix them up because the concepts are not that obvious in the different algorithms.

EXERCISES 4.3

Multiplying Fractions

Solve using a visual model and the standard algorithm.

1. $\frac{1}{3}$ of 18
2. $\frac{1}{3}$ of 36
3. $\frac{3}{4} \times 24$
4. $\frac{3}{8} \times 24$
5. $\frac{4}{5} \times 25$
6. $\frac{1}{7} \times 140$
7. $\frac{1}{2} \times 8$
8. $8 \times \frac{1}{2}$
9. $\frac{3}{5} \times 10$
10. $10 \times \frac{3}{5}$
11. $14 \times \frac{3}{7}$
12. $\frac{3}{4} \times 36$
13. $\frac{1}{2}$ of $\frac{2}{2}$
14. $\frac{2}{3}$ of $\frac{1}{2}$
15. $\frac{3}{4}$ of $\frac{4}{5}$
16. $\frac{2}{5}$ of $\frac{2}{3}$
17. $\frac{1}{2} \times \frac{3}{5}$
18. $\frac{2}{3} \times \frac{1}{4}$

Set up and solve with a visual model

19. $\frac{2}{3}$ of a number is 10. What's the number?
20. $\frac{3}{4}$ of a number is 24. What's the number?

Solve using the standard algorithm

21. $\frac{3}{4} \times \frac{2}{3}$
22. $\frac{4}{5} \times \frac{5}{8}$
23. $\frac{3}{4} \times \frac{5}{6}$
24. $\frac{2}{3} \times \frac{6}{7}$

25. $\frac{4}{9} \times \frac{3}{10}$ 26. $\frac{3}{11} \times \frac{7}{9}$ 27. $\frac{11}{12} \times \frac{13}{5}$

Applications of Multiplying Fractions. Solve using whatever technique you wish.

28. There are 48 students going on a field trip. One-fourth are girls. How many boys are going on the trip?

29. Abbie spent $\frac{5}{8}$ of her money and saved the rest. If she spent $45, how much money did she have at first?

30. A marching band is rehearsing in rectangular formation. $\frac{1}{5}$ of the marching band members play percussion instruments. $\frac{1}{2}$ of the percussionists play the snare drum. What fraction of all the band members play the snare drum?

31. Phillip's family traveled $\frac{3}{10}$ of the distance to his grandmother's house on Saturday. They traveled $\frac{4}{7}$ of the remaining distance on Sunday. What fraction of the total distance to his grandmother's house was traveled on Sunday?

32. Santino bought a $\frac{3}{4}$ pound bag of chocolate chips. He used $\frac{2}{3}$ of the bag while baking. How many pounds of chocolate chips did he use while baking?

33. Farmer Dave harvested his corn. He stored $\frac{5}{9}$ of his corn in one large silo and $\frac{3}{4}$ of the remaining corn in a small silo. The rest was taken to market to be sold.

a. What fraction of the corn was stored in the small silo?

b. If he harvested 18 tons of corn, how many tons did he take to market?

34. $\frac{5}{8}$ of the songs on Harrison's music player are hip-hop. $\frac{1}{3}$ of the remaining songs are rhythm and blues. What fraction of all the songs are rhythm and blues? Use a tape diagram to solve.

35. Three-fifths of the students in a room are girls. One-third of the girls have blond hair. One-half of the boys have brown hair.

a. What fraction of all the students are girls with blond hair?

b. What fraction of all the students are boys without brown hair?

4.4 DIVIDING FRACTIONS

Division is the process of finding the number used in the repeated additions of our related multiplication problem. Division is sometimes called the reciprocal relation of multiplication, and we will show why.

The division problem:

$$(\text{Number } 1) \div (\text{Number } 2)$$

is essentially asking the question; What number added to itself (Number 2) times equals (Number 1)?

Sometimes we rephrase this question and ask; How many (Number 2's) are there in (Number 1)?

If you think about this carefully you will see that both questions are actually asking you to find the same thing, but only in a different way. Why have two ways to ask the question? Because sometimes one question gives you a better grasp on understanding what is being asked for a particular problem than the other question. We'll show this through examples below.

We'd also like to point out that these questions are addressing what the concept of division really is. Eventually we will end up with a standard algorithm to use, but we don't want to lose sight of the concept of what we are actually doing. The algorithm is not division, it is just a set of steps used to quickly find our answer, with no obvious connection to the concept of division. We stress this point so that you don't get confused later on. Don't mistake the standard algorithm for division. Division is a concept, the standard algorithm is not.

We'll start by showing examples of division, for different types of division problems involving fractions.

Dividing a Fraction by a Whole Number

Let's start with the simplest form of division, that of dividing a fraction by a whole number. For this type of problem, the question:

What number added to itself (Number 2) times equals (Number 1)?

makes more sense intuitively. Anytime the fraction is the first number and a whole number is the second number you should use this conceptual interpretation of the division problem.

EXAMPLE 1: Perform the division $\frac{3}{4} \div 6$

We are actually asking; What number added to itself 6 times is equal to $\frac{3}{4}$? This question makes more sense than asking; How many 6's are in $\frac{3}{4}$? They both, however, are asking the same question.

Visual Model

To find the answer visually, let's look at the area model. We start with $\frac{3}{4}$

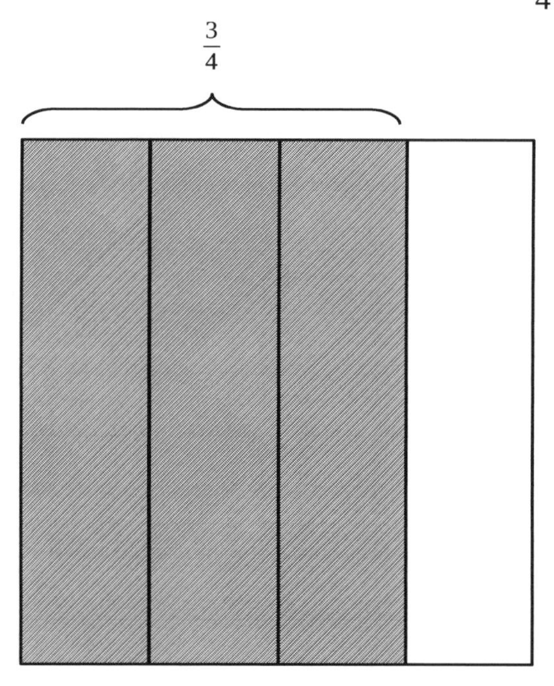

We then break (divide) it into 6 equal pieces, and figure out the size of an individual one sixth piece. We are looking for the amount in one part out of 6 of our 3/4, so we count how many pieces we have.

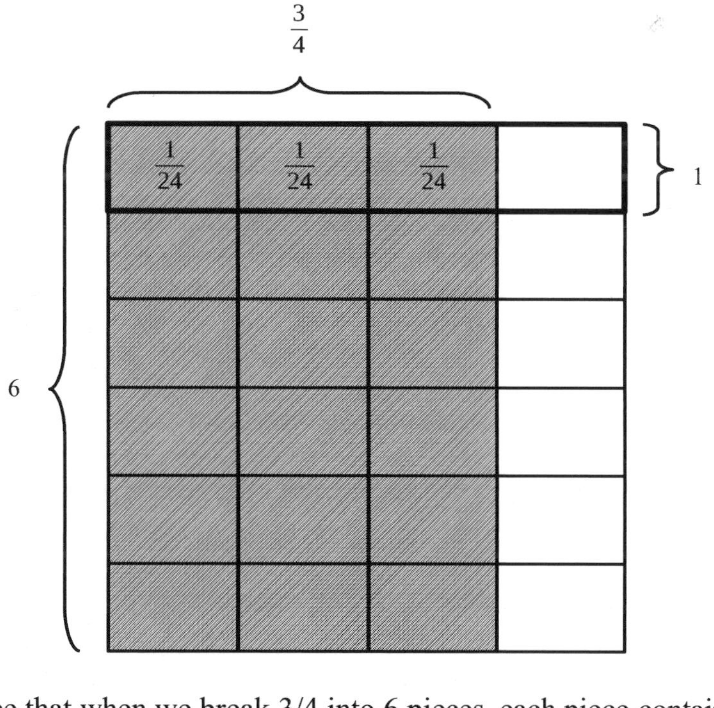

$\frac{3}{4}$

$\frac{1}{24}$ $\frac{1}{24}$ $\frac{1}{24}$ 1

6

From the figure we see that when we break 3/4 into 6 pieces, each piece contains boxes that are 1/24 in size. Then we simply count the number of 1/24 boxes we get, which is 3, so our final answer is.

$$\frac{3}{24} = \frac{1 \times 3}{8 \times 3} = \frac{1}{8}$$

The answer is $\frac{3}{24}$ or $\frac{1}{8}$. This means that if we add $\frac{3}{24}$ to itself 6 times we would have

$$\frac{18}{24} = \frac{3}{4}$$

We also see that the answer to the other question also works, but is harder to understand, i.e. there are $\frac{3}{24}$ 6's in 3/4.

Standard Algorithm

The standard algorithm is as follows. We start by changing all division problems to a multiplication by the reciprocal of the term we are dividing by. This makes sense for the problem above says that $\frac{3}{4} \div 6$ is equivalent to finding 1/6 of 3/4, or

$$\frac{3}{4} \div 6 = \frac{3}{4} \times \frac{1}{6}$$

Then we use the multiplication algorithm we developed earlier to obtain

$$\frac{3}{4} \times \frac{1}{6} = \frac{3 \times 1}{4 \times 6} = \frac{3}{24}.$$

Which we have shown is equivalent to $\frac{1}{8}$.

If we think about this, we can see that we were also looking for 1/6 of 3/4. This really is our multiplication problem in reverse. We want to know what number is 1/6 of 3/4 and that number is 1/8. This number is what we must add 6 times to get 3/4.

EXAMPLE 2:

Let's look at a slightly different example. Imagine that you have $\frac{2}{5}$ of a cup of frosting to share equally among three desserts. How would we write this as a division question? It is asking us to find $\frac{2}{5}$ of 3, or mathematically

$$\frac{2}{5} \div 3$$

Again, we are asking; What number added to itself 3 times is equal to $\frac{2}{5}$?

Visual Model

We can start by drawing a model of two-fifths.

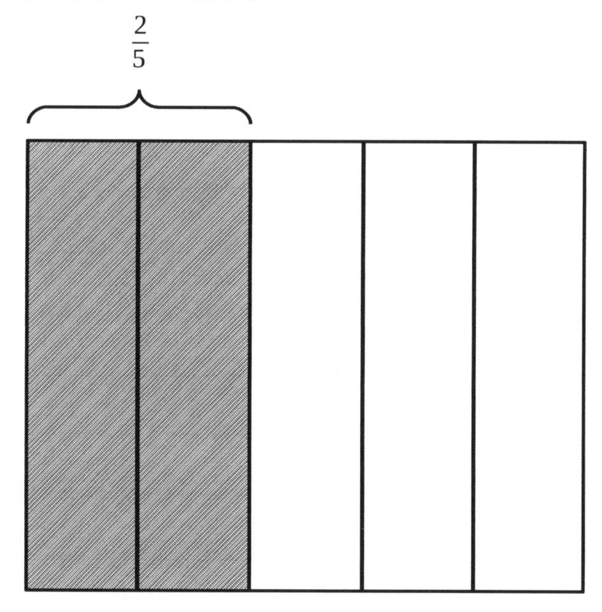

Next we show dividing two-fifths into three equal parts, and taking one of those 3 parts as our answer

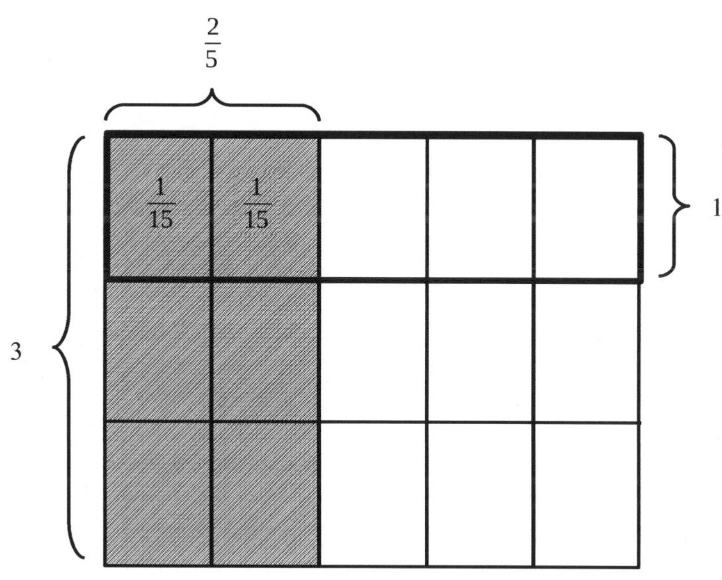

From the visual model, we can determine that $\frac{2}{5} \div 3 = \frac{2}{15}$.

Standard Algorithm

$$\frac{2}{5} \div 3 = \frac{2}{5} \times \frac{1}{3} = \frac{2 \times 1}{5 \times 3} = \frac{2}{15}$$

Dividing a Whole Number by a Fraction

For these type of problems we are asking the question; How many (Number 2's) are in (Number 1)? Anytime a whole number comes first it may make sense to phrase the division problem in this way. Again, it makes more sense intuitively for certain problems. This is due to our intuitive bias to think of a number as being a whole number and not a fraction. Sometimes the second question; What number added to itself (Number 2) times equals (Number 1), will make more sense, though. You should look at both so that you better understand what you are being asked to find.

EXAMPLE 3: Perform the division $5 \div \frac{1}{2}$

This problem asks the question; How many $\frac{1}{2}$'s are there in 5?

Visual Model

We start with 5 units

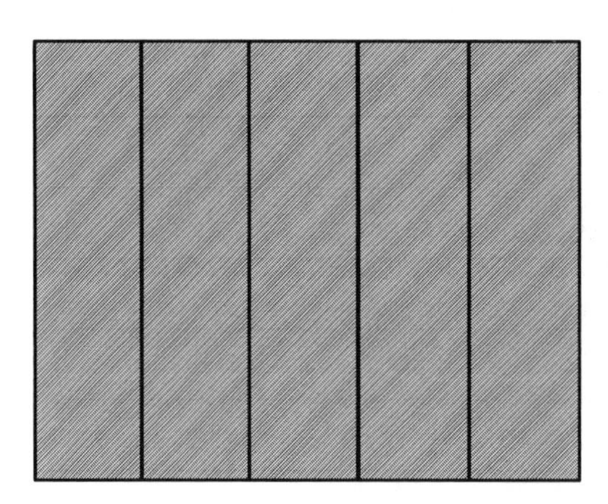

Divide it in half

$\frac{1}{2}$ $\frac{1}{2}$ $\frac{1}{2}$ $\frac{1}{2}$ $\frac{1}{2}$

$\frac{1}{2}$ $\frac{1}{2}$ $\frac{1}{2}$ $\frac{1}{2}$ $\frac{1}{2}$

And then count the number of half boxes.

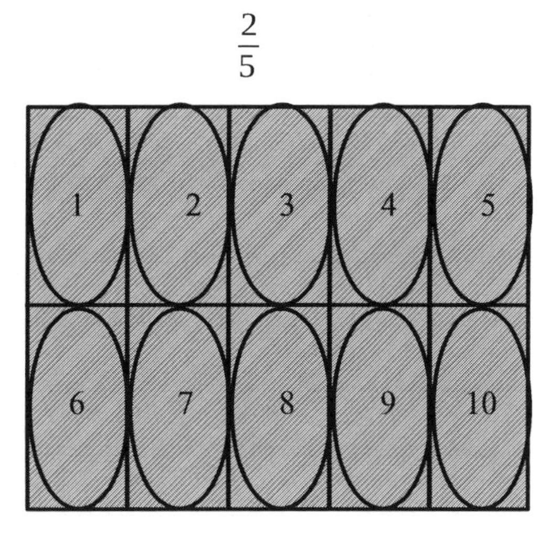

Our answer is $5 \div \frac{1}{2} = 10$

Standard Algorithm

Again, we just multiply 5 by the reciprocal of 1/2, which is 2.

$$5 \div \frac{1}{2} = 5 \times \frac{2}{1} = 10$$

EXAMPLE 4: Perform the division $3 \div \frac{2}{5}$

This problem asks the question; How many $\frac{2}{5}$'s are there in 3? or What number added to itself $\frac{2}{5}$ times equals 3. As you can see the second question is less intuitive.

Visual Model

We start with 3 units

Divide it into fifths and count how many two–fifths we have

$\frac{1}{5}$	$\frac{1}{5}$	$\frac{1}{5}$
$\frac{1}{5}$	$\frac{1}{5}$	$\frac{1}{5}$
$\frac{1}{5}$	$\frac{1}{5}$	$\frac{1}{5}$
$\frac{1}{5}$	$\frac{1}{5}$	$\frac{1}{5}$
$\frac{1}{5}$	$\frac{1}{5}$	$\frac{1}{5}$

We can see we have 7 and one left over. The answer is $7\frac{1}{2} = \frac{15}{2}$

Standard Algorithm

$$3 \div \frac{2}{5} = 3 \times \frac{5}{2} = \frac{3 \times 5}{2} = \frac{15}{2}$$

Dividing a Fraction by a Fraction

This is the final type of division problem we can expect. Since the second number is a fraction it makes more sense to ask the question;

How many (Number 2)'s are there in (Number 1)?

We start with an example where we divide a fraction by a fraction with the same denominator. We do this to develop an intuitive understanding before it becomes more abstract.

EXAMPLE 4: Perform the division $\frac{8}{9} \div \frac{2}{9}$

How many $\frac{2}{9}$'s are there in $\frac{8}{9}$?

Visual Model

We start with $\frac{8}{9}$ units

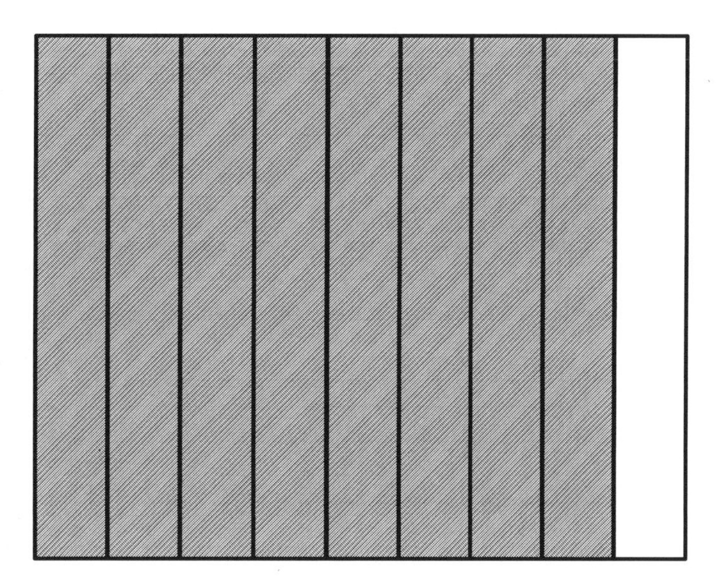

We then count how many $\frac{2}{9}$'s there are in this number.

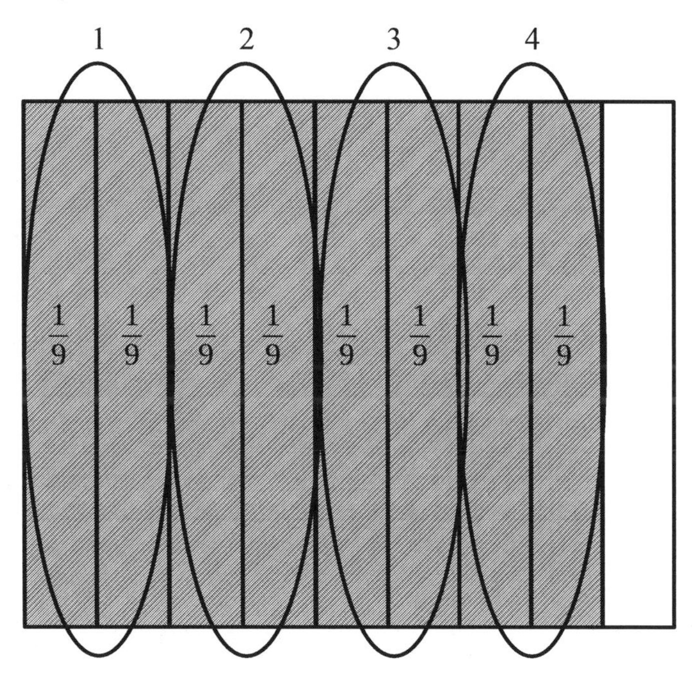

From our model we can see the answer is 4. Thus, $\frac{8}{9} \div \frac{2}{9} = 4$

<u>Standard Algorithm</u>

$$\frac{8}{9} \div \frac{2}{9} = \frac{8}{9} \times \frac{9}{2} = \frac{8 \times 9}{9 \times 2} = \frac{8}{2} = 4$$

EXAMPLE 5: Perform the division $\frac{9}{12} \div \frac{3}{12}$

How many $\frac{3}{12}$'s are there in $\frac{9}{12}$?

<u>Visual Model</u>

We start with $\frac{9}{12}$ units

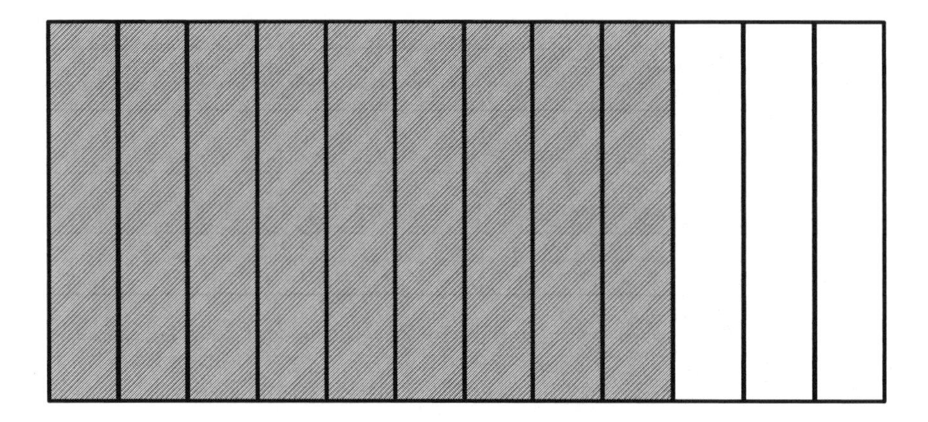

We then count how many $\frac{3}{12}$'s there are in this number.

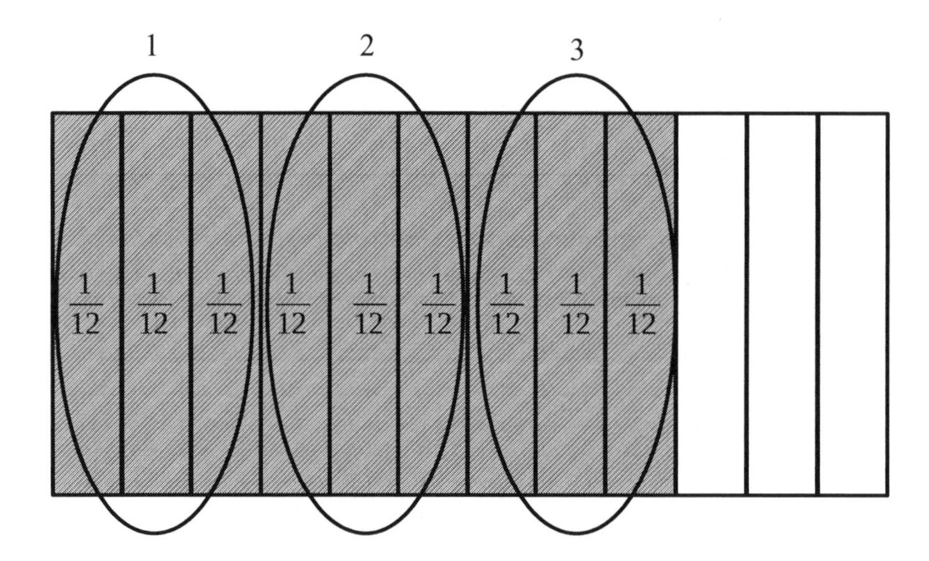

From our model we can see the answer is 3. Thus, $\frac{9}{12} \div \frac{3}{12} = 3$

Standard Algorithm

$$\frac{9}{12} \div \frac{3}{12} = \frac{9}{12} \times \frac{12}{3} = \frac{9 \times \cancel{12}}{\cancel{12} \times 3} = \frac{9}{3} = 3$$

We now do two examples where the denominators are not the same.

EXAMPLE 6: Perform the division $\frac{5}{6} \div \frac{3}{8}$

How many $\frac{3}{8}$'s are there in $\frac{5}{6}$?

Visual Model

We start with $\frac{5}{6}$

We need to count how many $\frac{3}{8}$'s there are in this number. To do this, we first need to construct $\frac{1}{8}$'s in our model. This can be a bit confusing, so bear with us.

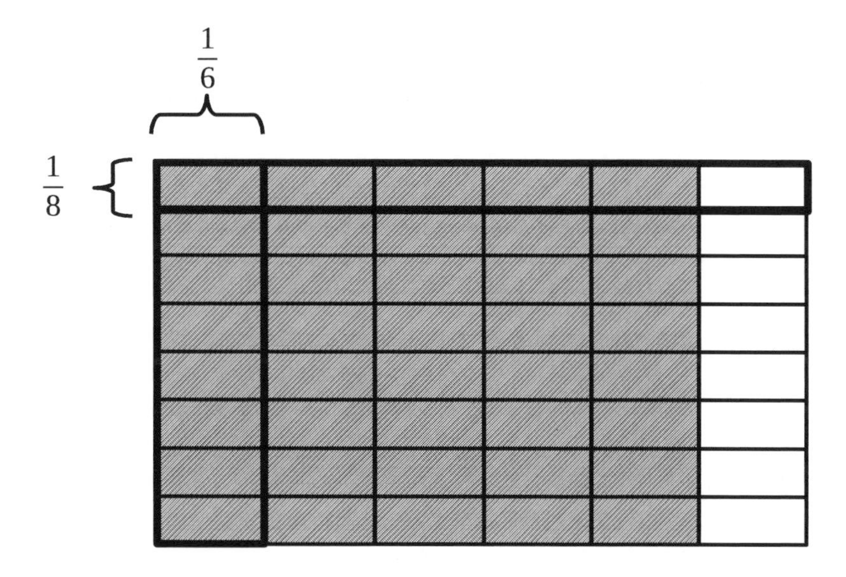

What we are showing here is $\frac{1}{8}$ and $\frac{1}{6}$ and where they overlap is actually $\frac{1}{48}$, but we don't really need this value. What we need to show is $\frac{3}{8}$ and count how many boxes of this size would be in a fully shaded $\frac{3}{8}$ amount. From what we see below, $\frac{3}{8}$ contains 18 of these boxes.

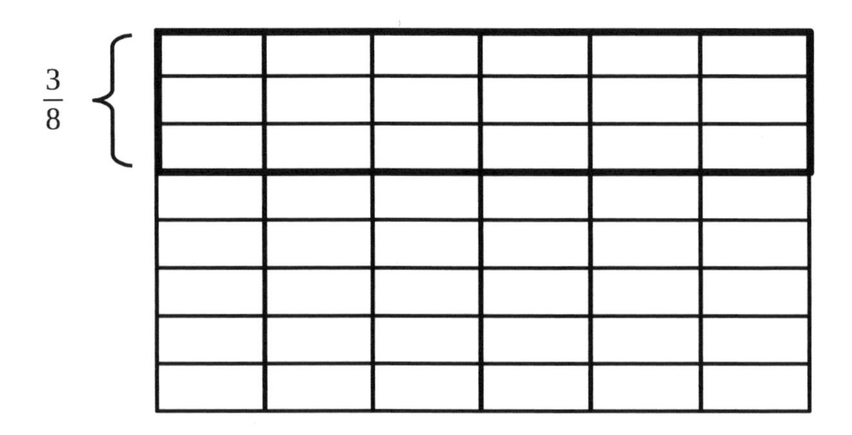

Thus we need to count 18 shaded boxes in our previous model, and then count how many groups of 18 can be formed.

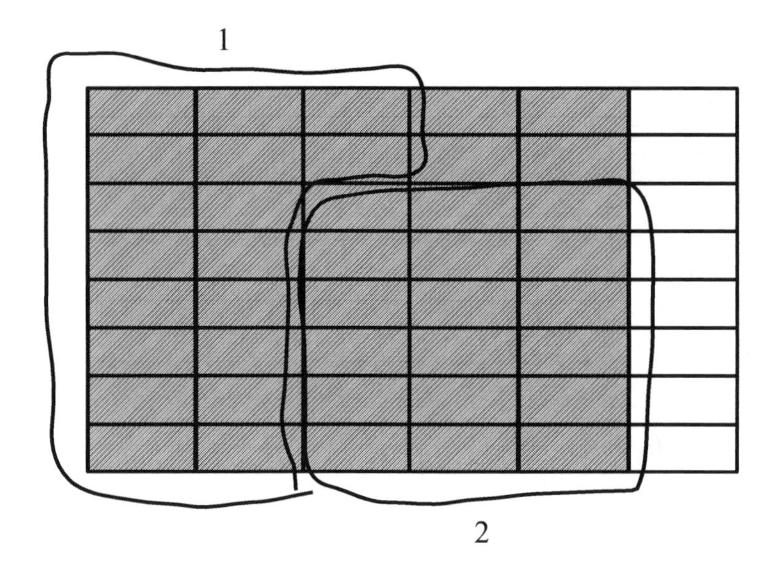

We see that we have two full groups and four left over. That means the last groups is 4 out of 18 or $\frac{4}{18}$. So, we have:

$\frac{5}{6} \div \frac{3}{8} = \frac{5}{6} \times \frac{8}{3} = \frac{5 \times 8}{6 \times 3} = \frac{40}{18} = \frac{20}{9}$, or $2\ \frac{4}{18} = 2\frac{2}{9}$ or we count the number of eighteenths to get $\frac{40}{18} = \frac{20}{9}$.

Standard Algorithm

$$\frac{5}{6} \div \frac{3}{8} = \frac{5}{6} \times \frac{8}{3} = \frac{5 \times 8}{6 \times 3} = \frac{40}{18} = \frac{20}{9}, \text{ or } 2\ \frac{4}{18} = 2\frac{2}{9}$$

Obviously, the standard algorithm is far less work and we don't want to do the models for every problem. We are simply using them here to demonstrate that there is a concept behind the division algorithm, but you really can't see it. That is the down side of an algorithm, it hides the concept, but it makes the process to get a result so much easier.

EXERCISES 4.4

Solve using a visual model to support your answer.

1. $\frac{3}{5} \div 6$
2. $\frac{2}{3} \div 4$
3. $\frac{5}{6} \div 3$
4. $\frac{4}{5} \div 2$
5. $4 \div \frac{2}{5}$
6. $3 \div \frac{2}{3}$
7. $7 \div \frac{3}{8}$
8. $9 \div \frac{1}{6}$
9. $\frac{4}{5} \div \frac{2}{5}$
10. $\frac{9}{4} \div \frac{3}{4}$
11. $\frac{7}{8} \div \frac{2}{8}$
12. $\frac{13}{10} \div \frac{2}{10}$
13. $\frac{11}{9} \div \frac{3}{9}$
14. $\frac{3}{4} \div \frac{2}{3}$
15. $\frac{3}{5} \div \frac{6}{8}$

Solve using the standard algorithm

16. $\frac{7}{9} \div \frac{8}{11}$
17. $\frac{5}{11} \div \frac{3}{7}$
18. $\frac{3}{7} \div \frac{5}{16}$
19. $\frac{11}{2} \div \frac{4}{3}$
20. $\frac{9}{13} \div \frac{3}{10}$
21. $\frac{6}{5} \div \frac{1}{7}$
22. $\frac{1}{9} \div \frac{2}{5}$
23. $\frac{12}{5} \div \frac{7}{5}$

Applications for division of fractions

24. Mother has 20 pounds of salad. She wants each person to have $\frac{1}{4}$ pounds of salad. How many people can she feed with the 20 pounds of salad that she has?

25. I bought a 10-lbs of sugar to make cakes. If each cake needs $\frac{3}{4}$ cups of sugar. Assume that the 10-lbs of sugar is approximately 20 cups, how many cakes can I make with the 10 -lbs of sugar?

26. A piece of ribbon $\frac{18}{5}$ m long is cut into 12 shorter pieces of equal length. What is the length of each short piece?

Chapter 4 Practice Test

Add the following fractions with the standard algorithm and visually show this process using a number line or tape diagram.

1. $\frac{5}{12}+\frac{1}{12}$

2. $\frac{2}{7}+\frac{5}{14}$

Add the following fractions with the standard algorithm and visually show the process using an area model.

3. $\frac{2}{9}+\frac{4}{5}$

4. $\frac{1}{6}+\frac{5}{8}$

Add the following fractions with the standard algorithm.

5. $\frac{5}{21}+\frac{1}{7}$

6. $\frac{5}{8}+\frac{7}{18}$

7. $\frac{9}{22}+1$

8. Jonathan read $\frac{3}{5}$ of a book on Saturday and then $\frac{1}{3}$ of the book on Sunday. How much of the book has he read?

Subtract the following fractions with the standard algorithm and visually show the process using a number line and tape diagram.

9. $\frac{5}{12}-\frac{1}{12}$

10. $\frac{3}{4}-\frac{5}{12}$

Subtract the following fractions with standard algorithm and visually show the process using an area model.

11. $\frac{5}{8}-\frac{2}{5}$

12. $\frac{5}{6}-\frac{3}{10}$

Subtract the following fractions with standard algorithm.

13. $\frac{3}{5}-\frac{4}{15}$

14. $\frac{9}{22}-\frac{5}{33}$

15. A student did $\frac{2}{7}$ of her homework on Saturday morning and $\frac{2}{5}$ in the afternoon. She said she would finish the remaining homework in the evening. What part of her homework does she have to do in the evening?

Solve using a visual model and standard algorithm.

16. $\frac{1}{4}$ of 36

17. $\frac{4}{5}$ of 20

18. $9\times\frac{2}{3}$

19. $\frac{2}{7}\times\frac{2}{3}$

Set up and solve with a visual model.

20. $\frac{2}{5}$ of a number is 14. What is the number?

Solve using standard algorithm.

21. $\frac{6}{7}\times\frac{2}{3}$

22. $\frac{3}{11}\times\frac{7}{9}$

23. There are 48 students going on a field trip. Two-thirds are girls. How many boys are going on the trip?

24. Samuel bought a $\frac{3}{4}$ pound bag of chocolate chips. He used $\frac{2}{3}$ of the bag while baking. How many pounds of chocolate chips did he use while baking?

25/ $\frac{5}{8}$ of the songs on Harrison's music player are hip-hop, $\frac{1}{3}$ of the remaining songs are thythm and blues. What fraction of all the songs are thythm and blues? Use a tape diagram to solve.

Solve using a visual method to support your answer.

26. $\frac{4}{5}\div 2$

27. $3\div\frac{2}{3}$

28. $\frac{13}{10}\div\frac{2}{10}$

Solve using the standard algorithm.

29. $\frac{3}{7}\div\frac{5}{16}$

30. $\frac{9}{13}\div\frac{3}{10}$

31. Mary has 20 pounds of salad. She wants each person to have $\frac{1}{4}$ pounds of salad. How many people can she feed with the 20 pounds of salad that she has?

32. A piece of ribbon $\frac{18}{5}$ meters long is cut into 12 shorter pieces of equal length. What is the length of each short piece?

CHAPTER 5

Decimal Fractions (Decimals)

In the previous two chapters we saw that fractions can be considered as numbers, since they enable us to measure in–between values. However, the notation that we used for fractions, did not make them appear to be actual numbers, but rather pairs of numbers used to represent parts of a whole. Then we add to that there were infinitely many ways to write the same numbers using equivalent fractions, and you are bound to be confused. Finally, the rules for arithmetic are really quite complicated, so that the standard algorithms can confuse even the brightest students. This is the major reason why fractions are so difficult for many students to understand well enough to be able to master them.

In this section we will show that fractions can be represented in an alternative way, so that they behave more like the whole numbers we are familiar with. Instead of being written using the fraction notion of having a specific number for the numerator and one for the denominator, where we can have infinitely many different ways of writing the same value, we can standardize what we use for the denominators and just vary the numerator. This is really just an extension of the place–value numeral system we developed in Chapter 1, and makes many of the operations on fractions easier to work with and understand. This new notation is called decimal fractions or decimals for short.

5.1 DECIMAL NOTATION

A decimal fraction, or decimal for short, is a way to represent numbers that are not whole numbers in a way that is similar to how we write whole numbers. We rely on the place-value form of our numbers and simply extend it to fractional values.

In decimals we only use unit fractions that are multiples of 10, such as

$$\frac{1}{10} \text{ or one-tenth, } \frac{1}{100} \text{ or one-hundredth, } \frac{1}{1000} \text{ or one-thousandth, etc.}$$

We remove the denominator from our representation, since we will always use the same denominators with our decimal fractions, and just the numerator term will change.

For example, to write the combined sum of fractions

$$\frac{3}{10}+\frac{5}{100}+\frac{7}{1000}$$

we simply write this in the following way.

First, we need to identify that the number we are writing is a fraction. For this we introduce the decimal point notation. Anytime we see a period in a number, we are to interpret whatever comes after the period to the right as a fraction less than one. The period is called a decimal point.

Finally, we only write the numerator terms of the fractions, since the denominators are already known. Thus, we can write the sum of the three fractions above as

$$0.357$$

Where it is to be understood that the period, now called the decimal point, means that the numbers following this point are all numerators for fractions whose denominators come in increasing powers of 10 as shown above.

Now, for numbers that have a whole part greater than one, we simply write the whole number part of the number to the left of the decimal point, and the fraction part to the right of the decimal.
As an example, the number 524.75 is understood to be

$$500 + 20 + 4 + \frac{7}{10} + \frac{5}{100}$$

Using this new notation, we can also see the connection to the place–values we used earlier.

Alternatively we could write this number as:

$$5\times100 + 2\times10 + 4\times1 + 7\times\frac{1}{10} + 5\times\frac{1}{100}$$

with the values after the times symbols being the place–values.

This means we can extend our place values to numbers to the right of the decimal point representing fractions with powers of 10 in their denominators. The place–values are now:

...	Hundreds	Tens	Ones	.	Tenths	Hundredths	Thousandths	...
...	100	10	1	.	1/10	1/100	1/1000	...

We now provide a few examples of writing decimals in expanded notation.

EXAMPLE 1: Write the decimal in expanded form, 0.59

$$0.59 = 5\times\frac{1}{10} + 9\times\frac{1}{100}$$

EXAMPLE 2: Write the decimal in expanded form, 1.703

$$1.703 = 1 + 7\times\frac{1}{10} + 0\times\frac{1}{100} + 3\times\frac{1}{1000}$$

EXAMPLE 3: Write the decimal in expanded form, 36.34

$$36.34 = 3\times10 + 6 + 3\times\frac{1}{10} + 4\times\frac{1}{100}$$

EXAMPLE 4: Write the decimal in expanded form, 237.4981

$$237.4981 = 2\times100 + 3\times10 + 7 + 4\times\frac{1}{10} + 9\times\frac{1}{100} + 8\times\frac{1}{1{,}000} + 1\times\frac{1}{10{,}000}$$

Another important skill when working with decimals is to be able to write any decimal in fraction form. Thus, instead of all the individual place values, we only need one fraction. We show this through some examples.

EXAMPLE 5: Write 8.91 as a decimal

We simply count how many places the decimal ends to the right of the decimal point. In this case it is two and that is the hundredths place. Thus, our fraction will be over 100. We then take our decimal digits without the decimal point and place them over 100, and we are done.

$$8.91 = \frac{891}{100}$$

EXAMPLE 6: Write 0.0378 as a decimal

We simply count how many places the decimal ends to the right of the decimal point. In this case it is four and that is the ten-thousandths place. Thus, our fraction will be over 10,000. We then take our decimal digits without the decimal point and place them over 10,000, and we are done.

$$0.0378 = \frac{378}{10{,}000}$$

EXAMPLE 7: Write 213.8 as a decimal

We simply count how many places the decimal ends to the right of the decimal point. In this case it is one and that is the tenths place. Thus, our fraction will be over 10. We then take our decimal digits without the decimal point and place them over 10, and we are done.

$$231.8 = \frac{2318}{10}$$

EXERCISES 5.1

Write the decimals in expanded form

1. 12.3
2. 25.397
3. 309.109
4. 0.057
5. 288.3
6. 12,289.75
7. 9,765.3
8. 17.123
9. 38.937
10. 21.1535
11. 9.2876
12. 0.035
13. 1.19
14. 16.0843
15. 25.0938
16. 11.002
17. 45.02023

Rewrite the decimals as fractions

18. 39.3
19. 0.397
20. 39.07
21. 0.00901
22. 4.04
23. 8.793
24. 389.1
25. 8.049
26. 16.2

5.2 COMPARING DECIMALS

A necessary skill when working with decimals is to determine their order and to be able to place them correctly on a number line. In this section we show how to determine if a decimal number is either less than (<) or greater than (>) another decimal number.

When comparing two decimal you need to compare them place–value by place–value starting with largest place–value from the left. If the digits in the same place–values are the same then the numbers are the same. If, however, you hit a place–value where one digit is less than the other similarly placed digit, then that decimal is less than the decimal you are comparing it to. For example:

$$21.3174 < 21.32$$

The numbers are the same in the 10's, one's and tenths, but differ in the hundredth's place, Since 2 is greater than 1, 21.32 is larger than 21.3174, even though the second number has more digits.

Consider the following examples.

EXAMPLE 1: Compare 103.576 to 3.578

In this case it is obvious that $103.576 > 3.578$ since $100 > 3$.

EXAMPLE 2: Compare 0.049 to 0.05

The numbers are the same up until the hundredth's place. Since 4 is less than 5 we see that $0.049 < 0.05$

EXAMPLE 3: Compare 5.07081 and 5.1

The numbers are the same up until the tenth's place. Since 0 is less than 1 we see that $5.07081 < 5.1$

EXAMPLE 4: Order the set of numbers and then place them correctly on a number line:

{4.9, 5.07081, 5.1, 5.01, 5.7003, 5.68, 5.09}

Following the previous examples we can order these numbers as follows:

$$4.9 < 5.01, 5.07081 < 5.09 < 5.1 < 5.68 < 5.7003$$

Putting them correctly on a number line we have:

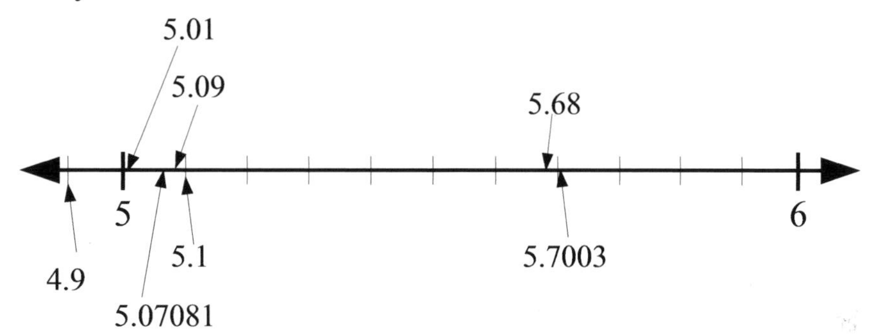

EXERCISES 5.2

Compare the decimals

1. 23.4, 23.39
2. 16.8, 16.09
3. 0.035, 0.03
4. 13.29, 3.3
5. 8.79, 7.8
6. 15.4309, 15.43

Order the set of numbers and then place them correctly on a number line:

7. {3.77, 3.02, 3.58, 3.95, 3.09, 3.8}
8. {0.23, 0.901, 0.3, 0.73, 0.703, 0.98}
9. {2.81, 3.0, 2.18, 2.06, 2.2, 2.005, 2.7, 2.71}
10. {11.9, 11.001, 11.71, 11.930, 11.554, 11.3, 11.06}
11. {1.7, 1.9, 1.07, 1.38, 1.83, 1.76, 1.67, 1.99, 2.0}
12. {5.05, 4.95, 5.15, 5.51, 5.85, 5.58, 5.976}

5.3 ADDING DECIMALS

You can add decimals in two different ways. The first approach turns the decimals into their place-value equivalent forms, and then you simply add the like fractions and whole numbers. If your fraction or whole number addition turns out to be larger than one, then you carry that whole number over to the next fraction in the place-value to the left.

EXAMPLE 1: Here is an example of adding 5.6 and 4.53 using the expanded fraction approach

Convert both to their place–value form:

$$5.6 = 5 + 6\times\frac{1}{10} \text{ and } 4.53 = 4 + 5\times\frac{1}{10} + 3\times\frac{1}{100}$$

Next we add the two and combine like place–value terms

$$\begin{aligned} 5.6 + 4.53 &= (5 + 4) + (5 + 6)\times\frac{1}{10} + (3 + 0)\times\frac{1}{100} \\ &= 9 + 11\times\frac{1}{10} + 3\times\frac{1}{100} \\ &= (9 + 1) + 1\times\frac{1}{10} + 3\times\frac{1}{100} \\ &= 10 + 1\times\frac{1}{10} + 3\times\frac{1}{100} \\ &= 10.13 \end{aligned}$$

EXAMPLE 2: Add 6.8 to 5.7 using the expanded fraction approach

$$\begin{aligned} 6.8 + 5.7 &= (6 + 5) + (8 + 7)\times\frac{1}{10} \\ &= 11 + 15\times\frac{1}{10} \\ &= (11 + 1) + 5\times\frac{1}{10} \\ &= 12 + 5\times\frac{1}{10} \\ &= 12.5 \end{aligned}$$

The second approach is called the standard algorithm for addition of decimals. This approach is the streamlined form of the first technique. It uses the same concepts, except that it organizes it a way that reduces what you have to write. The result is a nice algorithm, but it loses the connection to what you are actually adding,

The standard algorithm for adding decimals (along with the reasoning for it) is as follows:

1. Line up the decimal points of all the numbers you are adding (Guarantees you are adding like fractions / place-values)
2. Insert zeros so all decimals have the same number of digits (guarantees the fractions / place values all have digits to add)
3. Add the single digits in the decimals from right to left, carry any digits in the 10's place to the next decimal place to the left, and add with those digits during the next addition. Continue until you have passed all the way through the decimal digits. (Guarantees you are adding like whole numbers and fractions, and taking into account that when you have added enough to move to the next largest fraction / place-value in that decimal place you carry the result to that fraction / place-value).
4. Place the final decimal point at the same location as the other decimal points.

EXAMPLE 3: Add 5.6 and 4.53, using the standard algorithm

step 1) align decimal point $\begin{array}{r} 5.6 \\ +\ \underline{4.53} \end{array}$ step 2) insert zeros $\begin{array}{r} 5.60 \\ +\ \underline{4.53} \end{array}$

step 3) do single digit arithmetic with carrying $\begin{array}{r} \\ 5.60 \\ +\ \underline{4.53} \\ 3 \end{array}$, $\begin{array}{r} 1 \\ 5.60 \\ +\ \underline{4.53} \\ 13 \end{array}$, $\begin{array}{r} 1 \\ 5.60 \\ +\ \underline{4.53} \\ 10\ 13 \end{array}$

step 4) add decimal point $\begin{array}{r} 1 \\ 5.60 \\ +\ \underline{4.53} \\ 10.13 \end{array}$

Answer: 10.13

EXAMPLE 4: Add 13.9209 and 5.89, using the standard algorithm

step 1) align decimal point $\begin{array}{r} 13.9209 \\ +\ \underline{5.89} \end{array}$ step 2) insert zeros $\begin{array}{r} 13.9209 \\ +\ \underline{5.8900} \end{array}$

step 3) do single digit arithmetic with carrying $\begin{array}{r} \\ 13.9209 \\ +\ \underline{5.8900} \\ 9 \end{array}$, $\begin{array}{r} \\ 13.9209 \\ +\ \underline{5.8900} \\ 09 \end{array}$, $\begin{array}{r} 1 \\ 13.9209 \\ +\ \underline{5.8900} \\ 109 \end{array}$

$\begin{array}{r} 1\ 1 \\ 13.9209 \\ +\ \underline{5.8900} \\ 8109 \end{array}$, $\begin{array}{r} 1\ 1 \\ 13.9209 \\ +\ \underline{5.8900} \\ 9\ 8109 \end{array}$, $\begin{array}{r} 1\ 1 \\ 13.9209 \\ +\ \underline{5.8900} \\ 19\ 8109 \end{array}$

step 4) add decimal point $\begin{array}{r} 1\ 1 \\ 13.9209 \\ +\ \underline{5.8900} \\ 19.8109 \end{array}$

Answer: 19.8109

EXAMPLE 5: Add 32.3 to 75.083 and 16.0893, using the standard algorithm

step 1) align decimal point $\begin{array}{r} 32.3 \\ 75.083 \\ +\underline{16.0893} \end{array}$ step 2) insert zeros $\begin{array}{r} 32.3000 \\ 75.0830 \\ +\underline{16.0893} \end{array}$

```
                                                         1
                                              32.3000      32.3000
step 3) do single digit arithmetic with carrying  75.0830  ,  75.0830  ,
                                             +16.0893     +16.0893
                                                    3           23

      11            11            1  11          1  11
   32.3000       32.3000       32.3000        32.3000
   75.0830  ,    75.0830  ,    75.0830  ,     75.0830
  +16.0893      +16.0893      +16.0893       +16.0893
       723          4723        3 4723       123 4723
```

```
          1  11
          32.3000
step 4)   75.0830
         +16.0893
         123.4723
```

Answer: 123.4723

EXERCISES 5.3

Add the following decimals using the indicated technique.

Using expanded notation

1. 7.9 + 4.4
2. 8..6 + 6.1
3. 5.4 + 3.9
4. 11.7 + 7.9
5. 3.78 + 2.59
6. 13.6 + 7.35
7. 6.8 + 3.97
8. 24.36 + 16.9

Using the standard algorithm

9. 2.3 + 1.5
10. 21.3 + 17.2
11. 0.03 + 0.015
12. 15.7 + 6.8
13. 11.93 + 5.26
14. 14.3 + 8.65
15. 247.3 + 68.7
16. 774.06 + 123.5
17. 29.056 + 18.172
18. 37.33 + 17.54
19. 32.06 + 9.57
20. 16.2 + 3.89
21. 11.543 + 54.897
22. 0.395 + 2.785
23. 13.2 + 26.095
24. 8.07 + 12.387

5.4 SUBTRACTING DECIMALS

You can also subtract decimals in two different ways. The first approach turns the decimals into their place-value equivalent forms just like we did above for addition, and then you simply subtract the like fractions and whole numbers. If your fraction or whole number you are subtracting turns out to be larger than what you are subtracting from, then you must borrow from the larger number to the left before you can subtract.

EXAMPLE 1: Subtract 5.7 from 6.8 using the expanded fraction notation approach.

Convert both decimals to their expanded place-value form.

$$5.7 = 5 + 7\times\frac{1}{10} \text{ and } 6.8 = 6 + 8\times\frac{1}{10}$$

Subtract by combining like place-values terms.

$$6.8 - 5.7 = (6 - 5) + (8 - 7)\times\frac{1}{10}$$
$$= 1 + 1\times\frac{1}{10}$$
$$= 1.1$$

EXAMPLE 2: Subtract 4.53 from 5.6 or 5.6–4.53. Here we show the expanded fraction approach.

Convert both to their expanded place–value form:

$$5.6 = 5 + 6\times\frac{1}{10} \text{ and } 4.53 = 4 + 5\times\frac{1}{10} + 3\times\frac{1}{100}$$

Next we subtract the second number from the first and combine like place–value terms

$$5.6 - 4.53 = \left(5 + 6\times\frac{1}{10}\right) - \left(4 + 5\times\frac{1}{10} + 3\times\frac{1}{100}\right)$$
$$= 5 + 6\times\frac{1}{10} - 4 - 5\times\frac{1}{10} - 3\times\frac{1}{100}$$
$$5.6 - 4.53 = (5 - 4) + (6 - 5)\times\frac{1}{10} + (0 - 3)\times\frac{1}{100}$$
$$= (5 - 4) + (5 + 1 - 5)\times\frac{1}{10} + (0 - 3)\times\frac{1}{100}$$
$$= (5 - 4) + (5 - 5)\times\frac{1}{10} + 10\times\frac{1}{100} + (0 - 3)\times\frac{1}{100}$$
$$= (5 - 4) + (0)\times\frac{1}{10} + (10+0-3)\times\frac{1}{100}$$
$$= (5 - 4) + (0)\times\frac{1}{10} + 7\times\frac{1}{10}$$
$$= 1 + 0\times\frac{1}{10} + 7\times\frac{1}{100}$$
$$= 1.07$$

The second approach is called the standard algorithm for subtraction of decimals. Just like for addition this approach is the streamlined form of the first technique. It also uses the same subtraction concepts, except that it too organizes it a way that reduces what you have to write. Similarly, the result is a nice algorithm, but it too loses the connection to what you are actually subtracting,

The standard algorithm for subtracting decimals (along with the reasoning for it) is as follows:

1. Line up the decimal points of all the numbers you are subtracting (Guarantees you are subtracting like fractions / place-values)

2. Insert zeros so all decimals have the same number of digits (guarantees the fractions / place values all have digits to subtract from)
3. Subtract the single digits in the decimals from right to left, borrow from the next largest place over to the left if any of the subtractions cannot be done. This means you are trying to subtract a larger number from a smaller number. Continue until you have passed all the way through the decimal digits. (Guarantees you are subtracting like whole numbers and fractions, and taking into account that when you cannot subtract you borrow an amount from the next largest place-value, so that you can do the subtraction).
4. Place the final decimal point at the same location as the other decimal points.

EXAMPLE 3: Subtract 4.53 from 5.6, using the standard subtraction algorithm

step 1) align decimal point $\begin{array}{r} 5.6 \\ -\ \underline{4.53} \end{array}$ step 2) insert zeros $\begin{array}{r} 5.60 \\ -\ \underline{4.53} \end{array}$

step 3) do single digit arithmetic with borrowing $\begin{array}{r} {}^{5\,10} \\ 5.\not{6}0 \\ -\ \underline{4.53} \\ 7 \end{array}$, $\begin{array}{r} {}^{5\,10} \\ 5.\not{6}0 \\ -\ \underline{4.53} \\ 07 \end{array}$, $\begin{array}{r} {}^{5\,10} \\ 5.\not{6}0 \\ -\ \underline{4.53} \\ 1\ 07 \end{array}$

step 4) add decimal point $\begin{array}{r} {}^{5\,10} \\ 5.\not{6}0 \\ -\ \underline{4.53} \\ 1.07 \end{array}$

Answer: 1.07

EXAMPLE 4: Subtract 5.89 from 13.9209, using the standard subtraction algorithm

step 1) align decimal point $\begin{array}{r} 13.9209 \\ -\ \underline{5.89} \end{array}$ step 2) insert zeros $\begin{array}{r} 13.9209 \\ -\ \underline{5.8900} \end{array}$

step 3) do single digit arithmetic with borrowing $\begin{array}{r} 13.9209 \\ -\ \underline{5.8900} \\ 09 \end{array}$, $\begin{array}{r} {}^{8\,10} \\ 13.\not{9}209 \\ -\ \underline{5.8900} \\ 109 \end{array}$, $\begin{array}{r} {}^{8\,10} \\ 13.\not{9}209 \\ -\ \underline{5.8900} \\ 0109 \end{array}$, $\begin{array}{r} {}^{10\,8\,10} \\ \not{1}3.\not{9}209 \\ -\ \underline{5.8900} \\ 8\ 0109 \end{array}$

step 4) add decimal point $\begin{array}{r} {}^{10\,8\,10} \\ \not{1}3.\not{9}209 \\ -\ \underline{5.8900} \\ 8.0109 \end{array}$

Answer: 8.0109

EXERCISES 5.4

Subtract the following decimals using the indicated technique.

Using expanded notation

25. 7.9 – 4.4
26. 8..6 – 6.1
27. 5.4 – 3.9
28. 11.7 – 7.9
29. 3.78 – 2.59
30. 13.6 – 7.35
31. 6.8 – 3.97
32. 24.36 – 16.9

Using the standard algorithm

33. 2.3 – 1.5
34. 21.3 – 17.2
35. 0.03 – 0.015
36. 15.7 – 6.8
37. 11.93 – 5.26
38. 14.3 – 8.65
39. 247.3 – 68.7
40. 774.06 – 123.5
41. 29.056 – 18.172
42. 37.33 – 17.54

5.5 ROUNDING DECIMALS

Before we learn how to multiply and divide decimals, we will learn how to round these numbers to the nearest place-value indicated. This will enable us to compute approximate values when multiplying or especially dividing two decimals.

First, let's review the process for whole numbers. For example, if we want to round 238 to the nearest tens, we want to determine if 238 is closer to 230 or 240. 230 and 240 are called the benchmark numbers. Below is the line segment that shows the two benchmark numbers and the location of 238.

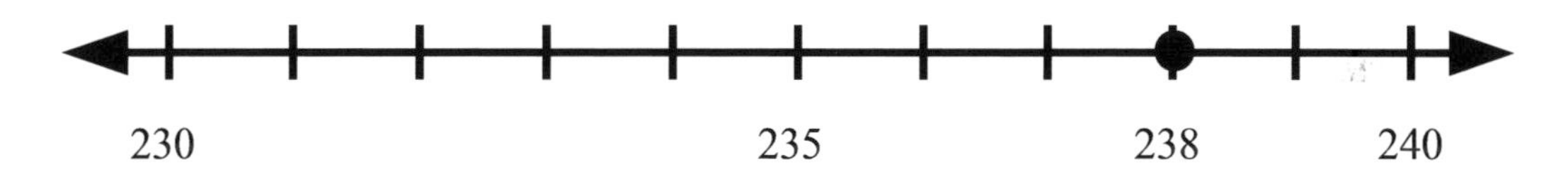

Notice that 235 is the midpoint between 230 and 240. We can see on the number line that 238 is closer to 240 or 24 tens, so we say that to round 238 to the nearest tens is 24 tens or 240.

Considering another example, we wish to round 2,318 to the nearest thousand, we first establish the benchmark numbers. They are 2,000 and 3,000.

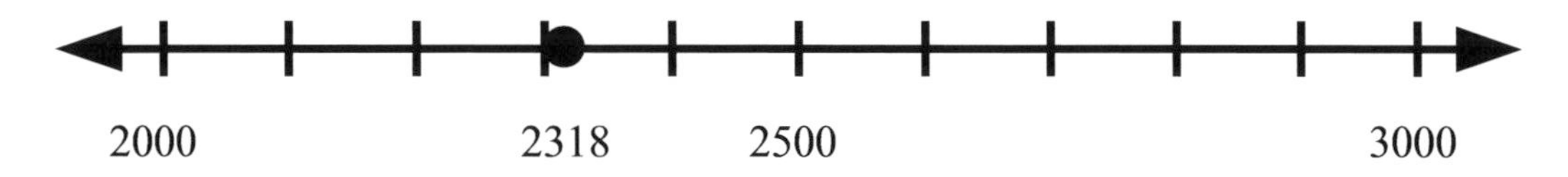

Since 2,318 is closer to 2,000, we say that 2,318 rounded to the nearest thousand is 2,000.

Suppose we want to round 2,318 to the nearest hundred, what would be the two benchmark numbers that we will use? They are 23 hundreds or 2,300 and 24 hundreds or 2,400.

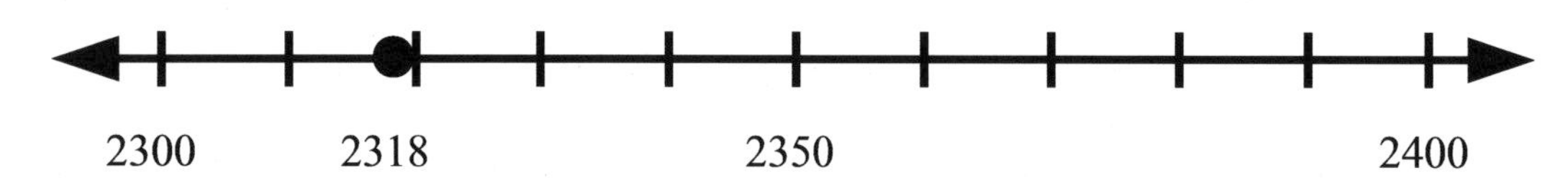

To do one more, we round 2,318 to the nearest tens, the two benchmark numbers are 2,310 and 2,320.

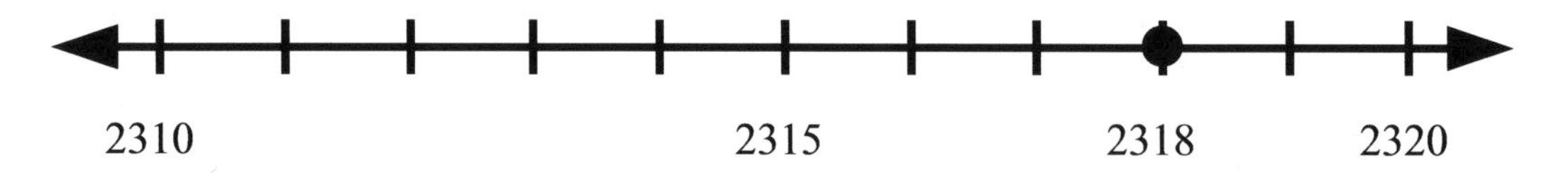

Since 2,318 is closer to 2,320 or 232 tens, we round 2,318 to the nearest tens which is 2,320.

It we want to round 2,315 to the nearest tens, we will adopt the rule to round it to the higher tens which in this case is 2,320. There ism another more complicated way to determine how to round the half-way mark between the benchmark numbers. But we will adopt the simpler rule.

Rounding decimal numbers to the nearest whole number, tenths, hundredths, and so on is done in the same way that we round the whole numbers. We figure out which two benchmark numbers the number we want to round are and use the number line to place the benchmark numbers. Then place the number we want to round on that number line and determine which benchmark number it is closer to.

EXAMPLE 1: Round 3.84 to the nearest tenths.

In this example the two benchmark numbers are 3.8 and 3.9.

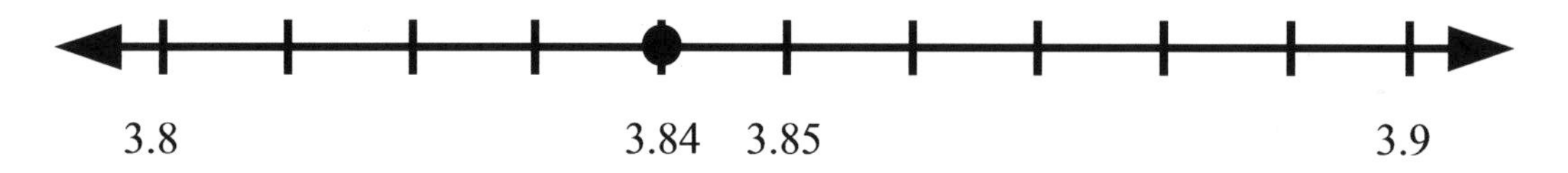

Since 3.84 is closer to 3.8, we round it to 3.8 or 3 and 8 tenths.

EXAMPLE 2: Round 3.84 to the nearest ones.

The two benchmark numbers are 3 and 4.

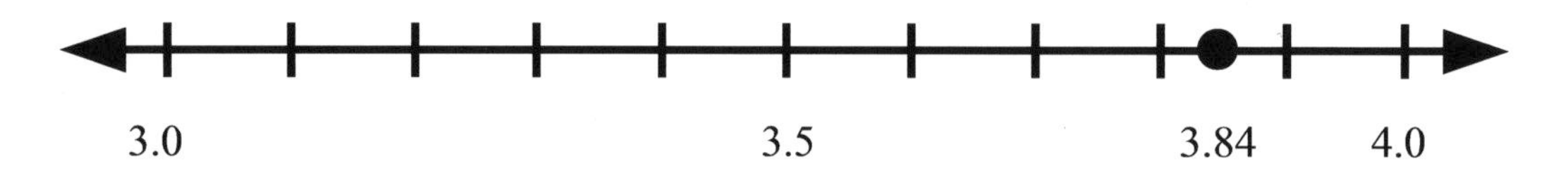

Since 3.84 is closer to 4, we round it to the nearest ones which is 4.

We could also do the rounding using a vertical line, but we'll stick to the horizontal number line.

EXAMPLE 3: Round 247.728 first to the nearest tenths and then to the nearest hundredths.

To round this number to the nearest tenths, the two benchmark numbers we will use are 247.7 and 247.8. The midpoint between these two numbers is 247.75.

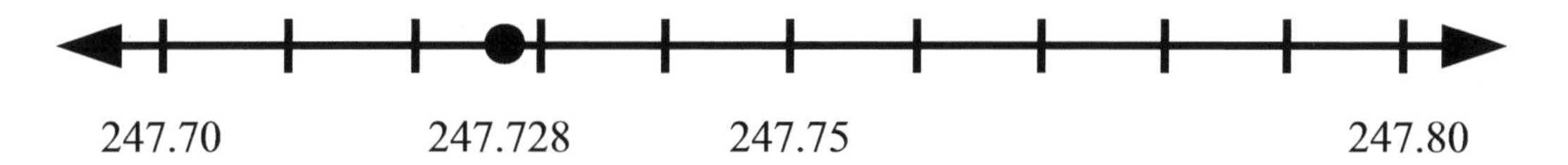

Since 247.728 is closer to 247.7, we round the number to 247.7.

Now round to the nearest hundredths. The two benchmark numbers are 247.72 and 247.73. The midpoint between these two numbers is 247.725.

7.728 is closer to 247.73, we round the number to 247.73.

EXAMPLE 4: Round 369.46 to the nearest tenths.

The two benchmark numbers that we use in this case are 369.4 and 369.5.

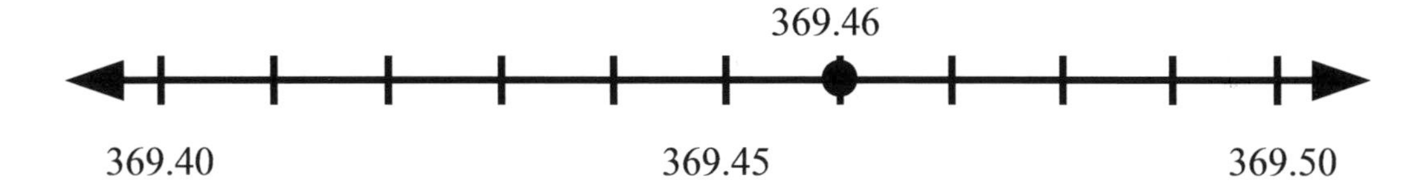

Since the number 369.46 is closer to 369.5, we round the number to its nearest tenths which is 369.5.

EXAMPLE 5: Round 369.46 to the nearest ones or whole number.

The two benchmark numbers in this case are 369 and 370.

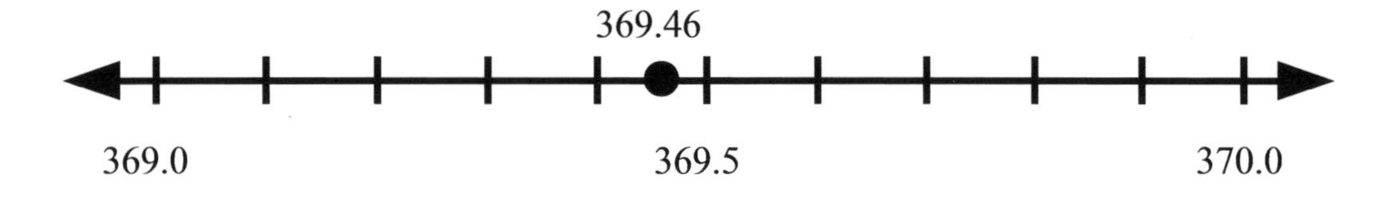

Since the number 369.46 is closer to 369, we round the number to the nearest whole number or ones which is 369.

EXERCISES 5.5

Round to the nearest whole number, tenth's, and hundredth's

13. 87.943
14. 389.854
15. 0.138
16. 2.793
17. 17.832
18. 20.097

19. Round the following numbers to the nearest tens.
 a) 376 b) 9,636 c) 126
20. Round the following numbers to the nearest hundreds.
 a) 3,786 b) 12,349 c) 9,023
21. Round the following numbers to the nearest tenths.
 a) 14.74 b) 6.472 c) 89.943
22. Round the following numbers to the nearest hundredths.
 a) 10.963 b) 59.095 c) 7.007
23. Round the following numbers to the nearest thousandths.
 a) 0.6718 b) 15.7093 c) 1.0086

5.6 MULTIPLYING DECIMALS

To multiply decimals we shall use the same approach we adopted in Chapter 4 for fractions. As you may recall, for fraction multiplication we showed that is was essentially similar to whole number multiplication. Instead of multiplying unit values we multiplied unit fractions. Thus, by treating decimals as fractions, we are implying that the process is actually equivalent to whole number multiplication. We illustrate this same approach for decimal using examples. There are three related approaches for doing this. The first uses the expanded fractional notation, and the second is the standard algorithm, while the third simply uses two fractions. We illustrate all of these in the examples below.

EXAMPLE 1: Multiply 3.2×2.4 using the expanded notation, the standard algorithm, and the multiplying fractions approaches.

In expanded notation,

$$3.2 = 3\times1 + 2\times \frac{1}{10} \text{ and } 2.4 = 2\times1 + 4\times \frac{1}{10}.$$

One of the first techniques we showed for multiplying whole numbers was the partial products approach from Chapter 1. Now, instead of just breaking up the whole numbers into their partial products, we must also include the decimal part of the place-value in the number, which we'll represent using the fraction notation.

We explain this as follows:

	2	$\frac{4}{10}$
3	6	$\frac{12}{10}$
$\frac{2}{10}$	$\frac{4}{10}$	$\frac{8}{100}$

In the table to the left we have written our two decimals using their expanded fraction form. One is written vertically and the other horizontally. Then we multiply the values in the various rows and columns and fill in the results as shown to the left. The answer is just the sum of these values in the bordered rectangles.

$$6+\frac{4}{10}+\frac{12}{10}+\frac{8}{100} = 6+\frac{16}{10}+\frac{8}{100}$$

$$= 6+\frac{10}{10}+\frac{6}{10}+\frac{8}{100} = 7+\frac{6}{10}+\frac{8}{100} = 7.68$$

This approach can also be used to illustrate what the standard algorithm for multiplying decimals is and why it works. This is show below:

```
  2.4
× 3.2
```

8	Multiply the 2 and the 4, which are really $\frac{2}{10}$ and $\frac{4}{10}$ to get $\frac{8}{100}$
4	Multiply the 2 and the 2, which are really 2 and $\frac{2}{10}$ to get $\frac{4}{10}$
12	Multiply the 3 and the 4, which are areally 3 and $\frac{4}{10}$ to get $\frac{12}{10}$
+ 6	Multiply the 3 and the 2, which is $3 \times 2 = 6$
7 68	Add the values vertically
7.68	Place the decimal point two spaces to the left

The last step regarding the placement of the decimal point has to do with the first place value from the right. Since we are multiplying a $\frac{2}{10}$ by a $\frac{4}{10}$ the results is $\frac{8}{100}$. This is why the 8 must be in the hundredths place so we must move over two place values. In the algorithm we always count how many places each decimal is to the right of the decimal point and add those values together to get how many places we move over in the final result.

We should also point out that when we multiplied the two 2's we moved the result, 4, to the left one place-value. This is because we are multiplying a "ones" times a "tenths" place-value, so the result should be one place value to the left of the 8. Similarly the 3 times the 4 is also placed with the right most number in the "tenths" place-value.

The approach just shown is not really the standard algorithm, but rather a detailed explanation of why it works. The standard algorithm is actually much simpler and can be streamlined even further. We just turn the process into whole number multiplication and properly place the decimal point.

The <u>standard algorithm for multiplying decimals</u> is as follows:

1. Line up the decimal points of all the numbers you are multiplying (Guarantees you are keeping track of the place-values you are multiplying)
2. Insert zeros so all decimals have the same number of digits and remove the decimal point (guarantees the fractions / place values all have digits to multiply and the correct place-value. Turns the process into whole number arithmetic)
3. Multiply the two whole numbers exactly like you would multiply two whole numbers.
4. Count the number of places to the right of each original decimal number after the zeros have been inserted, but before the decimal point was removed, and add these results
5. Place the final decimal point the number of places to the left by the value determined from the previous step.

Here is the streamlined version of the above example.

Step 1) $\begin{array}{r} 2.4 \\ \times\ 3.2 \\ \hline \end{array}$

Step 2) $\begin{array}{r} 24 \\ \times\ 32 \\ \hline \end{array}$

Step 3) $\begin{array}{r} 24 \\ \times\ 32 \\ \hline 48 \end{array}$, $\begin{array}{r} 24 \\ \times\ 32 \\ \hline 48 \\ 12\ \\ +\ 6\ \ \\ \hline 7\ 68 \end{array}$,

Step 4) 1 + 1 = 2 places

Step 5) $\begin{array}{r} 24 \\ \times\ 32 \\ \hline 48 \\ 12\ \\ +\ 6\ \ \\ \hline 7.68 \end{array}$

The third approach uses only fractions, and is show as follows.

First we convert each decimal into a fraction.

$$2.4 = \frac{24}{10} \text{ and } 3.2 = \frac{32}{10}$$

Then we multiply the fractions and perform the multiplication of fractions just as we did in Chapter 4. All these steps are shown below.

$$2.4 \times 3.2 = \frac{24}{10} \times \frac{32}{10} = \frac{24 \times 32}{10 \times 10} = \frac{768}{100} = 7.68$$

As you can hopefully see the process is relatively simple, provided we know how to transform decimals into fractions and multiply fractions.

EXAMPLE 2: Multiply 53.7×5.65 using the expanded notation, the standard algorithm, and the fraction approach.

In expanded notation, $53.7 = 5\times10 + 3\times1 + 7\times\frac{1}{10}$ and $5.65 = 5\times1 + 6\times\frac{1}{10} + 5\times\frac{1}{100}$.

Set up the partial products table.

	50	3	$\frac{7}{10}$
5	250	15	$\frac{35}{10}$
$\frac{6}{10}$	$\frac{300}{10}$	$\frac{18}{10}$	$\frac{42}{100}$
$\frac{5}{100}$	$\frac{250}{100}$	$\frac{15}{100}$	$\frac{35}{1000}$

In the table to the left we have written our two decimals using their expanded fraction form. One is written vertically and the other horizontally.

Then we multiplied the values in the various rows and columns and filled in the results as shown to the left. The answer is just the sum of these values in the bordered rectangles.

$$250+15+\frac{35}{10}+\frac{300}{10}+\frac{18}{10}+\frac{42}{100}+\frac{250}{100}+\frac{15}{100}+\frac{35}{1000}$$

$$= 250+15+3+\frac{5}{10}+30+1+\frac{8}{10}+\frac{4}{10}+\frac{2}{100}+2+\frac{5}{10}+\frac{1}{10}+\frac{5}{100}+\frac{3}{100}+\frac{5}{1000}$$

$$= (250+15+3+30+1+2)+\left(\frac{5}{10}+\frac{8}{10}+\frac{4}{10}+\frac{5}{10}+\frac{1}{10}\right)+\left(\frac{2}{100}+\frac{5}{100}+\frac{3}{100}\right)+\left(\frac{5}{1000}\right)$$

$$= (301)+\left(\frac{23}{10}\right)+\left(\frac{10}{100}\right)+\left(\frac{5}{1000}\right)$$

$$= 301+2+\left(\frac{3}{10}\right)+\left(\frac{1}{10}\right)+\left(\frac{5}{1000}\right)$$

$$= 303+\left(\frac{4}{10}\right)+\left(\frac{5}{1000}\right) = 303.405$$

Standard Multiplication Algorithm for Decimals

Step 1)
$$\begin{array}{r} 53.7 \\ \underline{\times\ 5.65} \end{array}$$

Step 2)
$$\begin{array}{r} 53.70 \\ \underline{\times\ 5.65} \end{array}, \quad \begin{array}{r} 5370 \\ \underline{\times\ 565} \end{array}$$

Step 3)
$$\begin{array}{r} 13 \\ 24 \\ 13 \\ 5370 \\ \underline{\times\ 565} \\ 26850 \\ 32220 \\ \underline{+26850} \\ 3034050 \end{array}$$

Step 4) 2 + 2 = 4 places

Step 5)
$$\begin{array}{r} 13 \\ 24 \\ 13 \\ 5370 \\ \underline{\times\ 565} \\ 26850 \\ 32220 \\ \underline{+26850} \\ 303.4050 \end{array}$$

Multiplying Fractions Approach

First we convert each decimal into a fraction.

$$53.7 = \frac{537}{10} \text{ and } 5.65 = \frac{565}{100}$$

Then we multiply the fractions and perform the multiplication of fractions just as we did in Chapter 4. All these steps are shown below.

$$56.7 \times 5.65 = \frac{537}{10} \times \frac{565}{100} = \frac{537 \times 565}{10 \times 100} = \frac{303{,}405}{1000} = 303.405$$

EXERCISES 5.6

Multiply the decimals using the partial product, standard algorithm, and fraction approaches

1. 3.0 × 5.4
2. 2.5 × 3.7
3. 7.4 × 2.1
4. 1.21 × 3.07
5. 8.561 × 2.3
6. 7.93 × 5.6
7. 3.6 × 7.38
8. 2.01 × 4.3
9. 0.35 × 4.2
10. 0.79 × 3.1
11. 0.05 × 0.103
12. 123.4 × 0.45

5.7 DIVIDING DECIMALS

To divide decimals we shall again use the same approach we adopted in Chapter 4 for fractions. As you may recall, for fraction division we showed that it essentially becomes multiplication by the reciprocal fraction. This will then transform the process into whole number arithmetic. We will introduce to techniques. The fraction approach and then the standard algorithm for dividing decimals. We illustrate all of these in the examples below

EXAMPLE 1: $16.2 \div 8.1$ using fraction notation, and the standard algorithm.

We transform the two decimals into fractions.

$$16.2 = \frac{162}{10} \text{ and } 8.1 = \frac{81}{10}.$$

Carry out the division of the fractions using the approach from Chapter 4.

$$16.2 \div 8.1 = \frac{162}{10} \div \frac{81}{10} = = \frac{162}{10} \times \frac{10}{81} = \frac{162 \times 10}{10 \times 81} = \frac{162 \times 10 \times 1}{81 \times 10}$$

$$= \frac{162 \times 10}{81} \times \frac{1}{10} = \frac{1620}{81} \times \frac{1}{10} = 20 \times \frac{1}{10} = 2$$

Notice in the calculations above how we chose to keep one factor of 10 in the numerator, but the other factor of 10 was moved to a new fraction of $\frac{1}{10}$ that we factored out. This will be the standard approach here. This enables us to do some basic whole number division, and then do the conversion back to the decimal in the end.

The standard algorithm for dividing decimals is as follows:

1. Rewrite the division problem using the long division symbol $\overline{)\quad}$
2. Remove the decimal points
3. Do normal long division of whole numbers.
4. Count the number of places to the right of each original decimal number. Subtract the number of places of the term you are dividing by from the number of places of the number you are dividing into it.
5. Place the final decimal point the number of places to the left by the value determined from the previous step.

Here is the standard division algorithm for this problem.

Step 1) $8.1\overline{)16.2}$ Step 2) $81\overline{)162}$ Step 3) $\begin{array}{r} 2 \\ 81\overline{)162} \\ \underline{-162} \\ 0 \end{array}$ Step 4) $1-1=0$

Step 5) Answer 2

EXAMPLE 2: 3.5 ÷ 1.25 using fraction notation, and the standard algorithm.

We transform the two decimals into fractions.

$$3.5 = \frac{35}{10} \text{ and } 1.25 = \frac{125}{100}.$$

Carry out the division of the fractions using the approach from Chapter 4.

$$3.5 \div 1.25 = \frac{35}{10} \div \frac{125}{100} = \; = \frac{35}{10} \times \frac{100}{125} = \frac{35 \times 100}{10 \times 125} = \frac{35 \times 100 \times 1}{125 \times 10}$$

$$= \frac{35 \times 100}{125} \times \frac{1}{10} = \frac{3500}{125} \times \frac{1}{10} = 28 \times \frac{1}{10} = 2.8$$

Notice in the calculations above how we chose to keep one factor of 10 in the numerator, but the other factor of 10 was moved to a new fraction of $\frac{1}{10}$ that we factored out. This will be the standard approach here. We will always keep at least one factor of 10 in the denominator. This enables us to do some basic whole number division, and then do the conversion back to the decimal in the end.

Here is the standard division algorithm.

Step 1) $1.25\overline{)3.5}$ Step 2) $125\overline{)350}$ Step 3)

$$\begin{array}{r} 28 \\ 125\overline{)350} \\ \underline{-250} \\ 1000 \\ \underline{-1000} \\ 0 \end{array}$$

Step 4) 2–1 = 1

Step 5) Answer 2.8

EXAMPLE 3: 6.077 ÷ 5.9 using fraction notation, and the standard algorithm.

We transform the two decimals into fractions.

$$6.077 = \frac{6077}{1000} \text{ and } 5.9 = \frac{59}{10}.$$

Carry out the division of the fractions using the approach from Chapter 4.

$$6.077 \div 5.9 = \frac{6077}{1000} \div \frac{59}{10} = \; = \frac{6077}{1000} \times \frac{10}{59} = \frac{6077 \times 10}{1000 \times 59} = \frac{6077 \times 10 \times 1}{59 \times 1000}$$

$$= \frac{6,077 \times 10}{59} \times \frac{1}{1000} = \frac{60,770}{59} \times \frac{1}{1000} = 1,030 \times \frac{1}{1000} = 1.03$$

Notice in the calculations above how we chose to keep one factor of 10 in the numerator, but the other factor of 1000 was moved to a new fraction of $\frac{1}{1000}$ that we factored out. This enables us to do some basic whole number division, and then do the conversion back to the decimal in the end.

The standard algorithm for dividing decimals approach is as follows:

Step 1) $5.9\overline{)6.077}$ Step 2) $59\overline{)6077}$ Step 3) $\begin{array}{r} 103 \\ 59\overline{)6077} \\ -59 \\ 177 \\ -177 \\ 0 \end{array}$ Step 4) 3–1 = 2

Step 5) Answer 1.03

An Important Note: In the previous examples the results of the division did not yield an infinitely repeating decimal. This was done on purpose. However, the overwhelming majority of problems will give a repeated decimal solution. We shall do one problem to show how this is handled. However, for these more complicated problems a calculator is a better option.

EXAMPLE 3: 3.5 ÷ 4.1 using fraction notation, and the standard algorithm.

We transform the two decimals into fractions.

$$3.5 = \frac{35}{10} \text{ and } 4.1 = \frac{41}{10}.$$

Carry out the division of the fractions using the approach from Chapter 4.

$$3.5 \div 4.1 = \frac{35}{10} \div \frac{41}{10} = = \frac{35}{10} \times \frac{10}{41} = \frac{35 \times 10}{10 \times 41} = \frac{35 \times 10 \times 1}{41 \times 10}$$

$$= \frac{35 \times 10}{41} \times \frac{1}{10} = \frac{350}{41} \times \frac{1}{10} =$$

We now notice that the division $\frac{350}{41}$ is not w whole number. In fact if we are to perform long division we will get $\frac{350}{41} = 8.5365\overline{85365}$. This is really beyond the scope of this course as it passes into some fairly complicated ideas relating to the concept of infinity. Thus, with this approach we can only yield a hybrid approach where we take the answer and divide it by 10 to obtain the final result $\frac{8.5365\overline{85365}}{10} = 0.85635\overline{85365}$.

The standard algorithm for dividing decimals approach is as follows:

Step 1) $4.1\overline{)3.5}$ Step 2) $41\overline{)35}$

$$\begin{array}{r} .853 \\ 41\overline{)\,35} \\ \underline{-328} \\ 22 \\ \text{Step 3)} \quad \underline{-205} \\ 15 \\ \underline{-123} \\ 27 \end{array}$$

etc.

This is an ongoing process until we see a repeating pattern. Thus, we need the more advanced Long Division Algorithm for infinitely many decimal places. Again, this is far beyond the scope of this course, but we wanted to at least show you a problem and point out is complexity. You will not see this on a test or the homework, unless you are allowed to use a calculator.

EXERCISES 5.7

Use fraction notation and the standard algorithm to divide the following.

1. $15.37 \div 2.9$
2. $11.625 \div 3.1$
3. $0.0136 \div 0.02$
4. $3.26625 \div 8.71$
5. $8.602 \div 2.3$
6. $0.0925 \div 3.7$
7. $6.804 \div 7.56$
8. $52.26 \div 7.8$
9. $14.4875 \div 4.75$
10. $105.5825 \div 13.45$
11. $46{,}421.01 \div 369.3$
12. $28{,}564.42 \div 98.6$

5.8 ESTIMATING WITH DECIMALS

Rounding numbers is used in estimating the operations of numbers.

EXAMPLE 1: Add 232 + 589.

We could easily add these two numbers using the standard algorithm. However, we could get a very quick estimate by rounding each of the addends to the nearest hundreds. 232 can be rounded to 200 while 589 can be rounded to 600 as it is closer to 600. Then we can add quickly 200 + 600 = 800 to see that our sum once we add these two numbers, it should be close to 800.
In fact when we add 232 + 589, we get 821.

EXAMPLE 2: Multiply 12.9×8.3.

We round 12.9 to ones which will be 13. We round 8.3 to the nearest ones which is 8.

Using 13×8, we get 104.

When we actually multiply the two numbers, 12.9×8.3, we get $\frac{129}{10} \times \frac{83}{10} = \frac{10{,}707}{100} = 107.07$

Note that the denominator is 100, that means the last digit of the product will be the hundredths.

We could round 12.9 to the nearest tens which is 10 and round 8.3 to the nearest tens too which is 10, the product $10\times10 = 100$ which is also not too far from the actual product value of 107.07.

Estimation is very useful particularly in multiplication and division of decimal numbers.

EXAMPLE 3: Multiply 32.82×79.23.

We could first round each number to the nearest tens which will allow us to do a quick calculation.

32.82 can be rounded to 30 as it is closer to 30 and
79.23 can be rounded to 80 as it is closer to 80.

Then we can multiply 30×80 which is 2400. Note that this is an estimation.

When we multiply these two numbers, Rewriting 32.82 as fraction, we have $32\frac{82}{100}$ which is $\frac{3282}{100}$ while 79.23 is $\frac{7923}{100}$.

Now, multiplying $\frac{3282}{100}\times\frac{7923}{100} = \frac{3282\times7923}{10{,}000} = \frac{26{,}003{,}286}{10{,}000} = 2{,}600.3286$

Note that the denominator is 10,000, so the last digit of the answer will be the ten thousandths.

Compare this answer from the estimate that we had which was 2,400.

EXERCISES 5.8

Find the decimal values of the following fractions and round to the nearest hundredths, if necessary:

42. $\frac{1}{9}$ 43. $\frac{2}{5}$ 44. $\frac{1}{7}$ 45. $\frac{1}{11}$ 46. $\frac{3}{8}$ 47. $\frac{1}{6}$

Estimate each of the following correct to 1 significant digit, then perform the operation using your calculator correct to 3 significant digits and compare your answer to the estimate.

1. 28.8×2.1
2. $52.3\div4.8$
3. 201.57×19.75
4. $201.57\div19.8$

Chapter 5 Practice Test

Rewrite the decimals in expanded notation.

1. 93,059
2. 21.325
3. 197.0134

Identify the place-value of the underlined digit.

4. $1\underline{7},038,769$
5. $102.03\underline{4}7$
6. $0.003\underline{9}5$

Compare the decimals

7. 16.8 , 16.09
8. 15.4309, 15.43

Order the set of numbers and then place them correctly on the number line:

9. {11.9, 11.002, 11.71, 11.930, 11.554, 11.3, 11.06}
10. {5.05, 4.97, 5.25, 5.85, 5.58, 5.967}

Add the following decimals using the expanded notation.

11. 8.6 + 6.1
12. 11.3 + 8.9
13. 23.46 + 16.9

Add the following decimals using the standard algorithm.

14. 23.6 + 19.4
15. 14.3 + 8.74
16. 37.33 + 15.74
17. 0.385 + 2.785

Subtract the following decimals using the expanded notation.

18. 8.6 – 6.1
19. 13.6 – 7.35
20. 24.36 – 18.9

Subtract the following decimals using standard algorithm.

21. 23.4 – 16.3
22. 14.1 – 8.75
23. 37.22 – 17.45

Round to the nearest whole number, tenths, and hundredths

24. 2.973
25. 389.845
26. 10.967

Multiply the decimals using the partial product, standard algorithm, and fraction approaches.

27. 1.23×3.07
28. 2.21×3.6
29. 312.4×0.46

Use fractional notation and the standard algorithm to divide the following.

30. $11.625 \div 3.1$
31. $52.26 \div 7.8$
32. $28{,}564.42 \div 98.6$

Find the decimal values of the following fractions and round to the nearest hundredths, if necessary.

33. $\frac{1}{7}$
34. $\frac{3}{8}$

Estimate each of the following correct to 1 significant digit, then perform the operation using your calculator correct to 4 significant digits and compare your answer to the estimate.

35. 312.4×18.2
36. $52.26 \div 7.8$

CHAPTER 6

Percents

6.1 WHAT ARE PERCENTS

In the previous sections we were introduced to the concept of decimals or decimal fractions. Fractions allow us to distinguish between parts of a whole, but as we have shown in previous sections, comparing fractions of different sizes can be quite complicated. In ancient Rome, long before the creation of decimals, parts out of a total of 100 became common. It was an easy way to separate small parts from a whole. Especially when taxes were levied. Taxes would typically be assigned in parts of a hundred, or in our modern fractional notation, $\frac{1}{100}$. In words this would be per one-hundred. In Latin one-hundred is a centum, so per one-hundred became percent (with centum shortened to cent and attached to the prefix per.) This new word was given a special symbol to identify it, and so we now use the % symbol.

So, when you take a test, and you receive a grade of 90%, what does it mean? It means that you scored 90 points out of 100 total points. The symbol % means out of 100 (per 100) and can also be written as $\frac{1}{100}$.

When we shop, we usually have to pay sales tax on the item that we purchased. For example, if the sales tax is 6%, then the question could be how much is the sales tax on a piece of furniture that I purchase. We see percents in many applications other than sales tax.

We can demonstrate this using the following visual model. The following is a 10 × 10 grid. To represent 90%, we can shade 90 of the unit boxes as shown below:

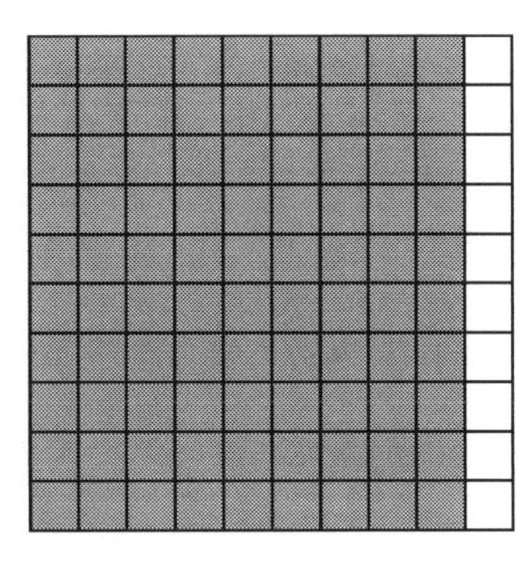

Suppose we have the following visual model, what percent does it represent?

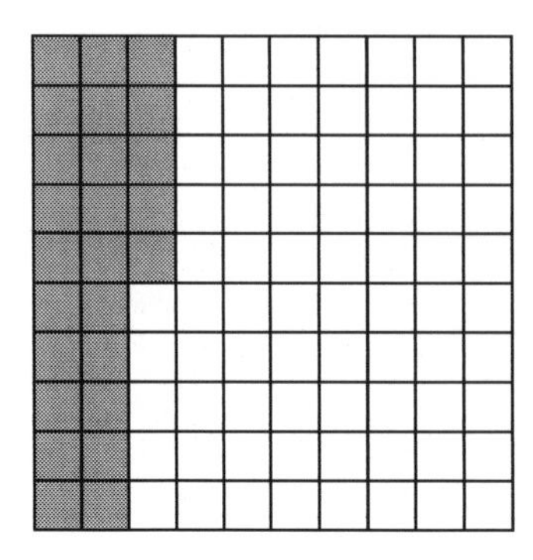

The shaded area is 25 boxes out of a total of 100 boxes, or $\frac{25}{100}$ which we call 25%. Here is another way of visually representing 25% as 1 out of 4 boxes, or $\frac{1}{4}$.

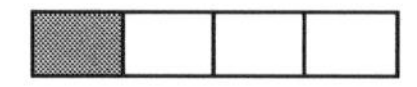

6.2 PERCENTS AS FRACTIONS AND DECIMALS

Percents can be written as a fraction or as a decimal. The example above represents 25% which can be written as either $\frac{25}{100}$ or $\frac{1}{4}$ or in decimal form as 0.25.

EXAMPLE 1: Write 63% in fraction form, in decimal form, and draw the visual model.

Fraction form: $63\% = 63 \times \frac{1}{100} = \frac{63}{100}$

Decimal form: $\frac{63}{100} = 0.63$

Visual model: Shade 63 out of 100 squares.

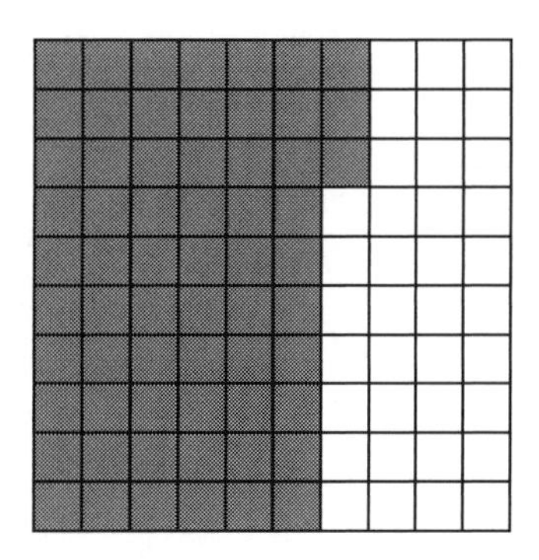

EXAMPLE 2: Write 22% as a fraction in reduced form.

$$22\% = \frac{22}{100} = \frac{2\times11}{2\times50} = \frac{\cancel{2}\times11}{\cancel{2}\times50} = \frac{11}{50}$$

EXAMPLE 3: Write 44.6% as a fraction in reduced form.

$$44.6\% = \frac{44.6}{100} = \frac{10\times44.6}{10\times100} = \frac{446}{1000} = \frac{2\times223}{2\times500} = \frac{\cancel{2}\times223}{\cancel{2}\times500} = \frac{223}{500}$$

EXAMPLE 4: Write $20\frac{1}{5}\%$ as a fraction in reduced form.

$$20\frac{1}{5}\% = (20+\frac{1}{5})\div100 = \frac{20}{100}+\frac{1}{5}\div100 = \frac{20}{100}+\frac{1}{5}\times\frac{1}{100}$$
$$= \frac{20}{100}+\frac{1}{500} = \frac{5\times20}{5\times100}+\frac{1}{500} = \frac{100}{500}+\frac{1}{500}$$
$$= \frac{101}{500}$$

EXAMPLE 5: Express 15% as a decimal

$$15\% = \frac{15}{100} = 0.15$$

EXAMPLE 6: Express 29.75% as a decimal

$$29.75\% = \frac{29.75}{100} = \frac{100\times29.75}{100\times100} = \frac{2,975}{10,000} = 0.2975$$

EXAMPLE 7: Express the fraction $\frac{13}{20}$ as a decimal and then as a percent.

$$\frac{13}{20} = \frac{5\times13}{5\times20} = \frac{65}{100} = 0.65$$
$$= \left(\frac{65}{100}\right)\times100\% = \left(\frac{65}{\cancel{100}}\right)\times\cancel{100}\% = 65\%$$

EXAMPLE 8: Express the decimal 0.45 as a percent then as a fraction in reduced form.

$$0.45 = 0.45\times100\% = 45\%$$
$$= \frac{45}{100} = \frac{5\times9}{5\times10} = \frac{\cancel{5}\times9}{\cancel{5}\times10} = \frac{9}{20}$$

EXAMPLE 9: Express the decimal 1.2 as a percent then as a fraction in reduced form.

$$1.2 = 1.2\times100\% = 120\%$$
$$= \frac{120}{100} = \frac{20\times6}{20\times5} = \frac{\cancel{20}\times6}{\cancel{20}\times5} = \frac{6}{5}$$

EXERCISES 6.2

Express each percentage as fractions in simplest form:

1. 19%
2. 50%
3. 38%
4. 90%
5. 62.5%
6. 125%
7. $30\frac{1}{2}$ %
8. $66\frac{2}{3}$ %

Express each of the following percents as decimals:

9. 50% 10. 30% 11. 27.5% 12. 21.25% 13. 45% 14. 62.5%

Express each fraction as a decimal and then as a percent:

15. $\frac{2}{5}$ 16. $\frac{7}{8}$ 17. $\frac{3}{40}$ 18. $\frac{12}{5}$

Express each decimal as a percent and then as a fraction in lowest terms:

19. 0.4 20. 0.38 21. 0.875 22. 1.92

6.3 APPLICATIONS OF PERCENTS

In this section we focus on three specific applications of percents. The first is finding the percentage of a given amount. The second has us trying to determine the original amount, given that we know the percent and the final amount. The third has us finding the percentage given the total and partial amounts. We introduce these through examples below.

EXAMPLE 1: What is 20% of 35?

$$20\%\text{ of }35 = \frac{20}{100}\times 35 = \frac{20}{100}\times\frac{35}{1} = \frac{20\times 35}{100\times 1} = \frac{700}{100} = \frac{7\times \cancel{100}}{1\times \cancel{100}} = 7$$

EXAMPLE 2: There are 450 students in a class, how many are in 25% of the class? You can use a calculator to approximate an answer.

$$25\%\text{ of }450 = 0.25\times 450 = 112.5$$

Since we cannot have a fraction of a student we round downwards and write the answer as approximately 112, or ≈ 112 students.

EXAMPLE 3: Adrienne's take home salary is $3,200 each month. She spends $1,300 for rent and $150 for gasoline for her car. What percent of her salary is her rent? What percent of her salary is for the gasoline she uses each month?

For rent we use 1,300 and 3,200 to obtain: $\frac{1300}{3200} = 0.40625 = 40.625\% \approx 41\%$

For gasoline we use 150 and 3,200 to obtain:

$$\frac{150}{3200} = 0.0046785 = 0.467855\% \approx 0.5\%$$

In addition to finding the percentage of a number given both the percentage and the number, we are sometimes given the percentage and the final value, and asked to find the number we are taking the percentage of only knowing the final value. Or given a part and the percent, can we find the whole. Sometimes we are given the part and the whole and are asked to find the associated percentage. Let's consider some examples to understand this better.

EXAMPLE 4: 60% of ***what number*** (the whole) is 300 (the part)?

This means we are looking for a number that when broken into 100 parts, 60 parts are equal to 300. Or equivalently, when that number is broken into 10 parts 6 of the parts add up to 300. From the tape diagram below, we can see this whole number must be 500. Since, breaking 500 into 10 parts gives 50 in each part, and 6 out of those 10 parts is equal to 300. So our answer is 500.

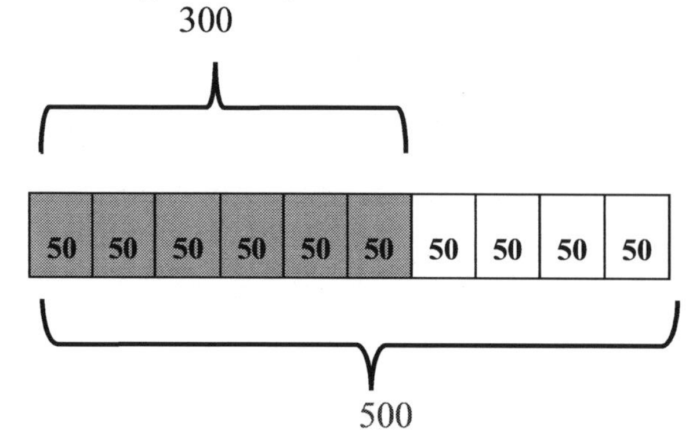

Or, $60\ \% = \dfrac{60 \times 5}{100 \times 5} = \dfrac{300}{500}$. Here, 300 is the part while 500 is the whole.

Now if we had to solve the problem using only this approach it would get quite cumbersome and difficult to do.

When we get to algebra, we will learn how to solve this type of problem using an equation, for now, we will divide 300 by 60% written as a fraction, or

$$\frac{300}{60\ \%} = \frac{300}{\dfrac{60}{100}} = 300 \div \frac{60}{100} = \frac{300 \times 100}{6} = \frac{300 \times 100}{60} = \frac{300 \times 100}{60} = 500$$

Let's consider another type of problem.

EXAMPLE 5: 60 out of 300 = what percent?

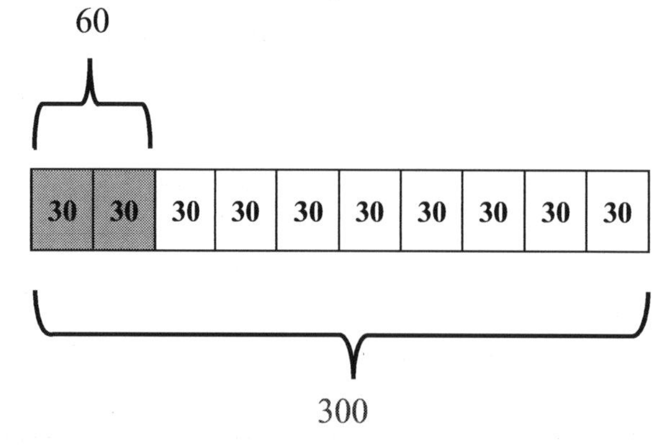

60 out of 300 can be written as $\frac{60}{300}$. Now divide both numerator and denominator by 3 to get 100 in the denominator:

$$\frac{60 \div 3}{300 \div 3} = \frac{20}{100}$$

So 60 out of 300 = 20%

Using decimals, we will divide 60 by 300 and multiply by 100%.

$$\frac{60}{300} \times \frac{100}{100} = 0.2 \times 100 \times \frac{1}{100} = 20 \times \frac{1}{100} = 20\%$$

EXERCISES 6.3

Applications of percents: (Use calculators as needed)

1. There are 600 students in a school. 35% of the students are girls. How many girls are in this school?

2. If 18 pieces of 50 pieces of candies are jolly ranchers. What percent does the 18 pieces represent?

3. Find the percent increase if a student's grade improved from 70 to 90.

4. Bernard's take home salary is $2000 each month. He spends $600 for rent and $100 for gasoline for his car. What percent of his salary is his rent? What percent of his salary is the gasoline he uses for each month?

Use either one of the two methods we showed above to do each of the following problems.
Use a visual model to show your work whenever possible.

5. What is 40% of 50?
6. What is 8% of 40?
7. What is 26% of 250?
8. What is 150% of 18?
9. 35 is 75% of what number?
10. 96% of what number is 432?
11. 160% of what number is 40?
12. 14 is 175% of what?
13. What percent of 70 is 21?
14. 37 is what percent of 148?
15. Suppose the sales tax is 8% and Mary purchased a sweater that costs $50, how much sales tax does Mary have to pay? When Mary goes to the cash register, what is the total amount that she has to pay including the sales tax?

16. There are 15 female and 25 male students in a class. What is the total number of students in the class? What percent of the class does the female students represent? What percent of the class does the male students represent?

17. Frank is real estate agent and assume that earns a 4% commission if he/she sells a house. Suppose he sold a house that costs $350,000, how much commission would Frank earn?

18. A company contributes 6% towards Elena's retirement 401K. Suppose Elena's gross salary is $1,500 per month, how much does the company contribute towards her 401K?

19. If 30% of the population in a town do not vote at last year's election, how many people did not vote last year if there were 6000 eligible voters in that town?

Chapter 6 Practice Test

Express each percentage as a fraction in simplest form:

1. 5%
2. 38%
3. 110%

Express each percent as a decimal:

4. 17%
5. 105%
6. 2%

Express each fraction as a decimal and then as a percent:

7. $\frac{4}{5}$
8. $\frac{3}{4}$

Express each decimal as a percent and then as a fraction in lowest terms:

9. 0.15
10. 0.04
11. 2.5

12. There are 400 students in a school. 40% of the students are girls. How many girls are in this school?

13. A company contributes 6% towards Eric's retirement 401K. Suppose Eric's gross salary is $2,500 per month, how much does the company contribute towards Eric's 401K?

14. Find the percent increase is a student's grade improved from 65 to 80.

15. What is 8% of 40?

16. What is 150% of 50?

17. 160% of what number is 48?

18. 40 is what percent of 120?

19. If there are 20 out of the 50 pieces of candy are coconut candy, what percent does the 20 pieces represent?

CHAPTER 7

Geometry, Measurement, Fractions and Decimals

In this chapter we focus on using fractions and decimals to measure geometric quantities. Whereas in Chapter 1 the focus was primarily on using whole numbers to count, in Chapter 2 we looked at whole numbers as a way to measure in geometry. In Chapters 3, 4 and 5 we introduced fractions, decimals and percents, so in this chapter we introduce the basics of geometry involving fractions and decimals, and show how to use these types of numbers to measure attributes of some additional geometric figures.

In addition we give an extended introduction to the idea of units presented at the beginning of this book. Our focus in this chapter is how to convert some common units to different scales for performing mathematical calculations. As we will shoe the process of converting one unit to anotehr unit relies on the fraction multiplication process. Another reason why knowledge of fraction arithmetic is important and indispensable.

7.1 MORE GEOMETRIC FORMULAS

In this section we introduce properties of geometric figures that require the use of fractions and decimals to calculate..

Object	**Figure**	**Formulas**
Triangle	a, c, h, b	Area = $\frac{1}{2}\times b\times h$

Object	Figure	Formulas
Circle	r	Circumference = $2\pi r$ Area = πr^2
Cylinder	r h	Volume = $\pi r^2 h$

EXAMPLE 1: Given a rectangle with length 6.3 in and width 3.5 in. find the perimeter and area of this rectangle.

6.3 in

3.5 in

Perimeter = 2L+2W = 2(6.3in) + 2(3.5in)
= 12.6in + 7in
= 19.6 in

Area = LW = (6.3in)(3.5in) = 22.05 in^2

EXAMPLE 2: Find the radius of a circle if its circumference is 33 cm. (Use the approximation for $\pi = 3.14$) Round your answer to the nearest hundredths.

Circumference $= 2\pi r = 33$cm

$$r = \frac{33\,\text{cm}}{2\pi} = \frac{33\,\text{cm}}{2(3.14)} = 5.25\,\text{cm}$$

The radius of the circle is approximately 5.25 cm.

EXAMPLE 3: Find the volume of a cylinder with radius 8.4 inches and height 11 inches. Use the approximation $\pi = 3.14$.

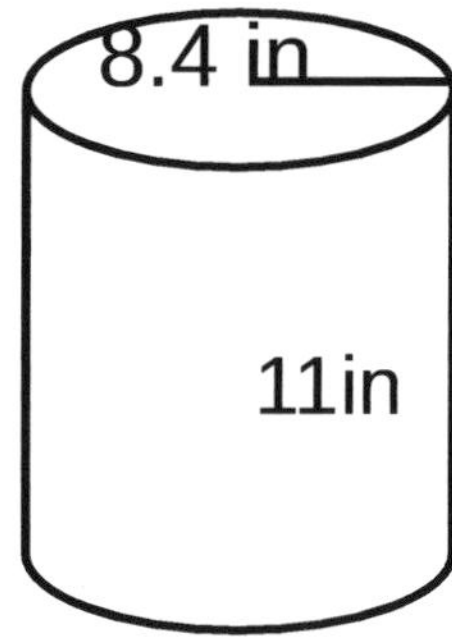

$$\text{Volume} = \pi r^2 h$$
$$= 3.14(8.4\text{ in})^2(11\text{in})$$
$$= 2437.1424\text{ in}^3$$

or 2,437.14 in^3 rounded to two decimal places.

EXERCISES 7.1

1. Find the perimeter of a square with a side that is 3.5 inches?

2. Find the perimeter of a rectangle that has a length of 4.6 cm and width of 2.7 cm.

3. Find the perimeter of a triangle that has sides of lengths 2.5 cm, 4.2 cm,and 3.8 cm.

4. Find the perimeter of a trapezoid if the sides of the parallel sides are 4.5 cm and 2.9 cm, and the two other sides are 5.1 cm and 4.7cm.

5. Find the area of a square with a side of 3.2 inches.

6. If the perimeter of a rectangular field is 68.4 m and the width of the field is 12.5 m, find the length of the field. Then find the area of the field.

7. The base of a triangle is 2.5 inches. The area of the triangle is 62.5 in^2. What is the height of the triangle?

8. The area of a square is 98 cm^2. Find the length of one side of the square. Use the approximation for $\pi = 3.14$ Then round your answer to the nearest tenths.

9. A box has dimensions of 2.4 ft by 5.3 ft. by10 ft. Find the volume of this box.

10. A box has dimensions of 4.4 ft by 3.3 ft. by 8 ft. Find the volume of this box.

11. A box has a base that is square of dimension of 3.5 in by 3.5 in. The volume of this box is 480 in^3. Find the height of box.

41. What is the perimeter of a $3\frac{1}{2}$ inch by 5 inch photo?

42. Bernie wants to put lights around his roof. Assume that his roof is rectangular and is $2\frac{1}{2}$ meters on one side and $9\frac{1}{4}$ meter on the longer side, how long should the lights be in order for Bernie to be able to put lights around his roof?

43. A triangle has sides $5\frac{2}{3}$ cm, $7\frac{2}{3}$ cm, and $9\frac{1}{3}$ cm. Find the perimeter of this triangle.

41. Find the perimeter of the following figure: Note that not all sides are given.

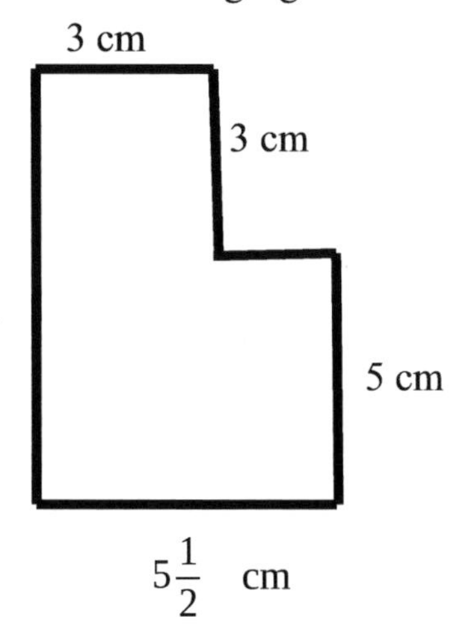

36. What is the area of a $5\frac{2}{3}$ cm by $3\frac{1}{3}$ inch photo?

37. What is the area of the following triangle?

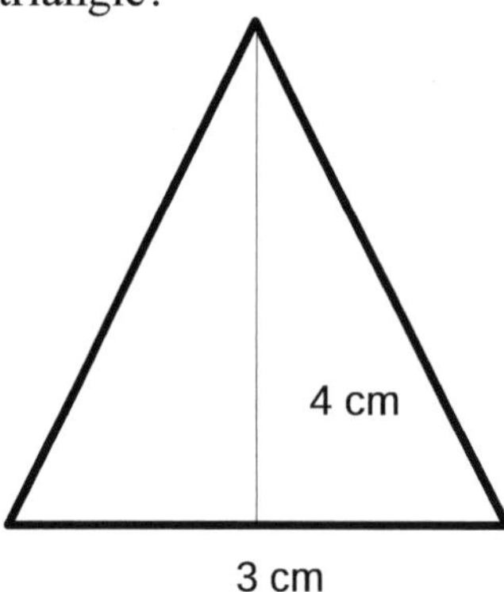

38. What is the area of a rectangle that has length of $10\frac{3}{4}$ cm and width of $6\frac{4}{5}$ cm?

24. The area of a triangular carpet is $66\frac{1}{2}$ m^2. If the height of the carpet is 3 m, what is the base of the triangular carpet? Give your answer in fraction form.
25. The area of a rectangular garden is 148 m^2. If the width of the garden is 20 m, what is the length? Leave your answer in fraction form.
26. The perimeter of a square kitchen is 18 m^2. Find the length of a side of the kitchen. Leave your answer in fraction form. Then find the area of the kitchen.
27. The perimeter of a rectangle is $48\frac{1}{3}$ cm. Its width is 6 cm. What is its length?

7.2 MEASUREMENT SCALES (SYSTEMS OF UNITS)

All measurements are relative and are made up of two distinct parts. The first part answers the how-much? question. It is the numerical quantification of the measurement such as 1, 2, 3, 6.5, or 10.75. The second part answers the "of what?" question. It identifies what we are measuring — it is the **units** part of our measurement. The units part, however, is also made up of two parts. The first is the particular attribute (unit) we are measuring such as length, mass, time, and temperature; and the second part is the scale we use to measure this attribute. In this case we have two distinct scales; English (also called Imperial) and metric.

Early on, we gave a very informal introduction of the different scales, simply by using them and relying on the fact that you have heard of them throughout your courses in mathematics and science in high school and earlier. We would now like to take a more complete look at the units we use and, in particular, the scales we use to measure them with.

Once we know "what" we are measuring, the next question is, "what scale" are we using to measure it? To be useful, a measurement must be comparable to some standard, i.e. some non-changing reference that we all agree upon and use. Over the years, trade and commerce between nations has driven us towards two common systems of measurements. The two systems are the English (or British) and the metric system. The metric system is by far the most dominant, due to its simplicity and the logical interrelationships between its units. The English System (no longer used in England) is what is primarily used in the US, and in just about NO other nation in the world. In fact, as of the writing of this book, there are only 3 nations that use the English system. They are the United States, Liberia and Myanmar. The rest of the ENTIRE world uses the metric system. Even in the US, the US Congress passed the Metric Conversion Act that then President Gerald R. Ford signed into law on December 23, 1975. It declared the metric system to be "the preferred system of weights and measures for United States trade and commerce", but permitted the use of United States customary units in non-business activities, yet we still follow the English system here.

Our primary focus in this book will be on the metric system, but for completeness we introduce the English system as well, mainly to see the two systems side-by-side, and to perhaps highlight why the rest of the world is using the more logical metric system approach. Additionally there is the practical reasons that we in the US still use this system, so we need to know how to calculate with it and understand how to use it.

The English System: An ad-hoc, confused, and sometimes humorous, system of measurements.

Though no longer used in England, many of the units in this system have been used for centuries and were originally based on common objects or human body parts, such as the yard and foot. We have all been trained extensively in the English system of measurement. Learning terms such as: inch, foot, yard and mile, or dry ounce, pound, and ton, or fluid ounce, pint, quart, and gallon. Additionally, there are all the crazy conversion factors needed to go from one to the other – twelve inches to a foot, three feet to a yard, 5,280 feet in a mile, sixteen dry ounces to a pound, 2,000 pounds to a ton, sixteen fluid ounces to a pint, two pints to a quart, four quarts to a gallon, 21, 24 or 44 gallons to a barrel. The list goes on and on.

Though practical when they were first created, the standards for the English units were not fixed and the standard could vary from object to object or from person to person. Eventually, the system was standardized, but the numerical relationship between common units in the system, as we'll see, was quite complicated.

In the English system the first four of the seven fundamental units are:

Quantity	English Base Unit
Length	foot (ft)
Mass	pound (lb)
Time	second (s)
Temperature	Fahrenheit (°F)

From these units many different types of units are derived. As you can see from all the different multiplication factors discussed above, the English system of measurement is quite complicated. In the next section with give some common unit conversions that you should be familiar with.

The Metric System: A system based upon logic and reason

Since our real number system is all based upon powers of ten, why not have our measurement system conform to the same pattern? This way we won't have all these crazy numbers to memorize to try and change from one measure to another. This is quite logical, and makes sense. That is the basis of the metric system. It is just like our base–10 (place–values with powers of 10) number system.

Most people don't want to change over to the metric systems, because they haven't been trained to **think** in metric. Over the years we have learned to get a sense for how big an inch, foot, and mile. are, or how cold 28°F is, and we are comfortable with these quantities. However, if you ask the average person to perform calculations between the various units, mistakes are inevitable, because the conversions, as shown earlier, are quite complicated and appear to be without a rational reason.

We need to be trained to think in metric units, so that we can then easily compute using the metric system, and move away from the English system. Here are some ways to think-metric:

- A parking meter, the height of a six year old boy, four basketballs placed side by side, a baseball bat, 4 pieces of dry spaghetti laid end to end, the distance from an adult man's nose to his finger tips when he puts his arms out sideways, seven one dollar bills placed end to end, a stack of $100 in dimes, are all about a meter in length.

- The width of a small fingernail, or the width of a black key on the piano, is about the size of a centimeter (one one-hundredths of a meter).

- The width of the ball on a ballpoint pen, thickness of a dime, paperclip, or credit card, are about the size of a millimeter (one one-thousandths of a meter).

- The weight of a small paperclip, or a kidney bean is about a gram.

Once we establish a comfortable point of reference, it is so much easier to do calculations using the metric system.

The first four of the seven fundamental units for the metric system are:

Quantity	Metric Base Unit
Length	meter (m)
Mass	kilogram (kg)
Time	second (s)
Temperature	Kelvin (K), or Celsius (^{o}C)

The larger and smaller versions of each unit in the metric system do not require a series of complicated conversion tables to represent, as we needed for the English system. Instead, the related measurements are created either by multiplying, or dividing, the base unit by factors of ten. This feature provides a great convenience to users of the system, as it eliminates the need for such cumbersome calculations as dividing by 16 (to convert ounces to pounds) or by 12 (to convert inches to feet). Similar calculations in the metric system are performed simply by shifting the decimal point in the numerical part of the measurement. The metric system is a base-10 or decimal system. How these units are subdivided or multiplied, is through a series of prefixes that specify the size relative to the standard measure for each of the fundamental units: meter, gram and liter.

Size of Units and Prefixes

If you want to measure the distance between cities, the meter might be considered too small. You wouldn't measure the distance from New York to Los Angeles in meters any more than you would measure it in yards or feet. You really need a larger unit of measure. Similarly, you wouldn't use a meter stick to measure distances at the cellular or atomic level. You really need a much smaller unit of measure. To accommodate very large and very small measures, a series of word prefixes are used to signify multiples or fractions of the fundamental meter unit. To simplify matters, all the related larger units are multiplied by factors of 10 and all the smaller units are divided by multiples of 10. Some commonly used prefixes with their meanings and numerical equivalences are:

Prefix	Meaning	Abbreviation	Number	Exponent Form
Tera-	trillion	T	1,000,000,000,000	10^{12}
Giga-	billion	G	1,000,000,000	10^{9}
Mega-	million	M	1,000,000	10^{6}
kilo-	thousand	k	1,000	10^{3}
centi-	hundredths of	c	0.01	10^{-2}
milli-	thousandths of	m	0.001	10^{-3}
micro-	millionths of	μ	0.000001	10^{-6}
nano-	billionths of	n	0.000000001	10^{-9}
pico-	trillionths of	p	0.000000000001	10^{-12}

Prefixes within the same units are related to one another. There are 1000 millimeters in a meter, 100 centimeters in a meter, and 1000 meters in a kilometer, i.e.

100 centimeters = 1 meter
1000 meters = 1 kilometer

A little less obvious, but something that can be shown is:

10 millimeters = 1 centimeter

Whenever we make a measurement, we need to pick the appropriate prefix depending upon the size of the unit we are measuring. For example, the correct metric unit for distances between cities is a kilometer (km). That prefix, kilo-, means that we're talking about a unit that is 1,000 times larger than the basic unit of a meter. Thus a kilometer is 1,000 times the length of one meter. If you know how many meters it is between New York and Portland, it's very easy to figure out the distance in kilometers (divide by 1,000). That is much easier than with English units. In the metric system, the conversions only involve factors of ten.

The scientific community has adopted the metric system and has agreed upon a set of standards called the International Standard or SI system of units. Since just about all the countries of the world have adopted this standard, we need to be familiar with the two different measurement systems – the English and the metric systems. However, in the world of science the metric system is the system of choice, and the system that every scientist should be intimately familiar with.

7.3 CONVERTING UNITS

In this section we learn how to convert units, both within a single system, as well as converting units between the English and the metric systems. We introduce the approach by using something called **dimensional analysis**. Dimensional analysis is a process where we convert a problem by using only conversion factors to obtain the final unit type. We start by understanding what the beginning unit is and then we need to determine the sequence of multiplication factors to obtain the desired unit after all the unit (dimension) cancellations. We illustrate this approach through examples.

Converting Units Within a Single System

The process of converting in either the English or the metric is the same. The only difference is that for the English system the conversion factors are varied, while for the metric they are all powers of 10. We'll start with the metric system, as this is easier than the English system.

Conversions Within the Metric System

To convert similar type units within the metric system we must first identify the size of the unit we are converting from, and the size of the unit we wish to convert to. This will tell us what conversion factors to use. Next, we choose the conversion factor that will convert our unit, e.g. meter, gram or liter. We then determine the conversion factor that will convert the meter, gram or liter unit into our desired unit. We then multiply by the appropriate conversion factors and write the new unit designation.

A conversion factor is a ratio of two equivalent physical quantities expressed in different units. For example, we know that 1,000 mm = 1 m. If we divide both sides by 1,000 mm we have:

$$\frac{1{,}000\text{ mm}}{1{,}000\text{ mm}} = \frac{1\text{ m}}{1{,}000\text{ mm}} \text{ or } 1 = \frac{1\text{ m}}{1{,}000\text{ mm}}$$

The conversion factor is $\frac{1\text{ m}}{1{,}000\text{ mm}}$ and, as we have shown, this equals 1. We can continue in the same way and derive conversion factors between the units and derive the following conversion tables below:

Converting to meter (also gram, or liter): "From" the unit in the denominator

From Unit	To Unit	Conversion Factor
mm mg mL	m g L	$\frac{1(\text{m},\text{g},\text{L})}{1{,}000\text{ m}(\text{m},\text{g or L})}$
cm cg cL	m g L	$\frac{1(\text{m},\text{g},\text{L})}{100\text{ c}(\text{m},\text{g or L})}$

From Unit	To Unit	Conversion Factor
mm mg mL	cm cg cL	$\frac{1\,c(m,g,L)}{10\,m(m,g\text{ or }L)}$
km kg kL	m g L	$\frac{1{,}000\,(m,g,L)}{1\,k(m,g\text{ or }L)}$

Converting from meter (also gram, or liter): "To" the unit in the numerator

From Unit	To Unit	Conversion Factor
m g L	mm mg mL	$\frac{1{,}000\,m(m,g\text{ or }L)}{1(m,g,L)}$
m g L	cm cg cL	$\frac{100\,c(m,g\text{ or }L)}{1(m,g,L)}$
cm cg cL	mm mg mL	$\frac{10\,m(m,g,L)}{1\,c(m,g\text{or }L)}$
m g L	km kg kL	$\frac{1\,k(m,g\text{ or }L)}{1{,}000\,(m,g,L)}$

We illustrate the process with several examples:

EXAMPLE 1: Convert 15 mm into m.

We are converting from mm to m. We identify the appropriate conversion factor from the table above as $\frac{1\text{ m}}{1{,}000\text{mm}}$, since we are converting from mm to m. **The "from" unit is always in the denominator.** Now multiply and cancel the units as follows:

$$15\text{ mm} = \frac{15\text{ mm}}{1} \cdot \frac{\text{m}}{1{,}000\text{ mm}} = 0.015\text{ m}$$

Notice that the mm units cancel to leave the desired m units. This is what dimensional analysis is. We choose the conversion factor so that the wanted dimension does not cancel, but the other units do. This, as you can see, is identical to how we multiplied fractions in Chapter 1. If a quantity in is both the numerator and denominator (top and bottom of the expression) it cancels out.

EXAMPLE 2: Convert 32.5 cg into mg.

We are converting from cg to mg. In using the table, we will need to use the cg to mg conversion factor. Thus, using dimensional analysis, we go from cg to mg.

We identify the appropriate conversion factor from the table above as $\frac{10\,mg}{1\,cg}$. We obtain this simply by replacing the m for meter in the above table with g for gram everywhere, since we are converting from cg to cg,

Now multiply and cancel the units as follows:

$$32.5\,cg = \frac{32.5\,\cancel{cg}}{1} \cdot \frac{10\,mg}{1\,\cancel{cg}} = 325\,mg$$

Notice that the cg units cancel to leave the mg units.

EXAMPLE 3: Convert 0.75 L into mL.

We are converting from L to mL. We identify the appropriate conversion factor from the table above as, $\frac{1{,}000\ mL}{1L}$ since we are converting from L to mL. The "from" unit is always in the denominator. Now multiply and cancel the units as follows:

$$0.75\ L = \frac{0.75\ \cancel{L}}{1} \cdot \frac{1{,}000\ mL}{1\ \cancel{L}} = 750\ mL$$

Notice that the L units cancel to leave the mL units.

Conversions Within the English System

We now show how to convert within the English system. We derive the conversion factors in the same way we have already shown for the metric conversions. This leads to the following conversion tables:

Convert larger to smaller units:

Convert "from" "to"	Conversion Factor
feet to inches	$\frac{12\ \text{inches}}{1\ \text{foot}}$
yard to feet	$\frac{3\ \text{feet}}{1\ \text{yard}}$
miles to feet	$\frac{5280\ \text{feet}}{1\ \text{mile}}$
pint to ounces	$\frac{16\ \text{ounces}}{1\ \text{pint}}$

Convert "from" "to"	Conversion Factor
quart to pints	$\frac{2 \text{ pints}}{1 \text{ quart}}$
gallon to quarts	$\frac{4 \text{ quarts}}{1 \text{ gallon}}$
pounds to ounces	$\frac{16 \text{ ounces}}{1 \text{ pound}}$
tons to pounds	$\frac{2000 \text{ pounds}}{1 \text{ ton}}$

Convert smaller to larger units:

Convert "from" "to"	Conversion Factor
inches to feet	$\frac{1 \text{ foot}}{12 \text{ inches}}$
feet to yards	$\frac{1 \text{ yard}}{3 \text{ feet}}$
feet to miles	$\frac{1 \text{ mile}}{5280 \text{ feet}}$
ounces to pints	$\frac{1 \text{ pint}}{16 \text{ ounces}}$
pints to quarts	$\frac{1 \text{ quart}}{2 \text{ pints}}$
quarts to gallons	$\frac{1 \text{ gallon}}{4 \text{ quarts}}$
ounces to pounds	$\frac{1 \text{ pound}}{16 \text{ ounces}}$
pounds to tons	$\frac{1 \text{ ton}}{2000 \text{ pounds}}$

EXAMPLE 1: Convert 72 feet to yards.

We are converting from feet to yards. We identify the appropriate conversion factor from the table above as $\frac{1 \text{ yard}}{3 \text{ feet}}$. Now multiply and cancel the units as follows:

$$\begin{aligned} 72 \text{ feet} &= \frac{72 \text{ feet}}{1} \cdot \frac{1 \text{ yard}}{3 \text{ feet}} \\ &= \frac{72 \text{ yards}}{3} \\ &= 24 \text{ yards} \end{aligned}$$

Notice that the feet units canceled to leave the yard units.

EXAMPLE 2: Convert 2.5 miles to feet.

We are converting from miles to feet. We identify the appropriate conversion factor from the table above as $\frac{5280 \text{ feet}}{1 \text{ mile}}$, Now multiply and cancel the units as follows:

$$\begin{aligned} 2.5 \text{ miles} &= \frac{2.5 \text{ miles}}{1} \cdot \frac{5280 \text{ feet}}{1 \text{ mile}} \\ &= \frac{2.5(5280 \text{ feet})}{1} \\ &= 13{,}200 \text{ feet} \end{aligned}$$

Notice that the mile units canceled to leave the feet units.

EXAMPLE 3: Convert 468 ounces to gallons.

We are converting from ounces to gallons. Using dimensional analysis, we must first convert ounces to pints, then pints to quarts. and finally quarts to gallons.

We identify the appropriate conversion factors from the tables above as $\frac{1 \text{ pint}}{16 \text{ ounces}}$, $\frac{1 \text{ quart}}{2 \text{ pints}}$, and $\frac{1 \text{ gallon}}{4 \text{ quarts}}$.

Now multiply and cancel the units as follows:

$$\begin{aligned} 468 \quad \text{ounces} &= \frac{468 \quad \text{ounces}}{1} \cdot \frac{1 \quad \text{pint}}{16 \quad \text{ounces}} \cdot \frac{1 \quad \text{quart}}{2 \quad \text{pints}} \cdot \frac{1 \quad \text{gallon}}{4 \quad \text{quarts}} \\ &= \frac{468 \quad \text{gallons}}{(16)\,(2)\,(4)} \\ &= \frac{468 \quad \text{gallons}}{128} \\ &\approx 3.7 \quad \text{gallons} \end{aligned}$$

Notice that the ounces, pints, and quarts units canceled to leave the gallons units.

EXAMPLE 4: Convert 1.25 tons to pounds (lbs).

We are converting from tons to pounds. We identify the appropriate conversion factor from the table above as

$$\frac{2000 \text{ pounds}}{1 \text{ ton}}$$

Now multiply and cancel the units as follows:

$$1.25\,\text{tons} = \frac{1.25\,\cancel{\text{tons}}}{1} \cdot \frac{2000 \text{ lbs}}{1\,\cancel{\text{ton}}} = 2{,}500\,\text{lbs}$$

Notice that the tons units cancel to leave the lbs units.

Converting Between the English and Metric Systems

This tends to be the most complicated, since it involves many different types of units. For this we will derive two conversion tables. The first is metric to English, and the second is English to metric. The conversion factors will be obtained as was done previously. Since $1 \text{ meter} = 3.28 \text{ feet}$ we divide both sides by 1 meter to obtain the conversion factor from meters to feet as:

$$\frac{1 \text{ meter}}{1 \text{ meter}} = \frac{3.28 \text{ feet}}{1 \text{ meter}}$$

$$1 = \frac{3.28 \text{ feet}}{1 \text{ meter}}$$

We can continue in this way to derive the conversion tables:

English to Metric

Convert "from" "to"	Conversion Factor
inch to cm	$\frac{2.54 \text{ cm}}{1 \text{ inch}}$
feet to meter	$\frac{1 \text{ m}}{3.28 \text{ feet}}$
yard to meter	$\frac{0.9144 \text{ m}}{1 \text{ yd}}$
ounce to mL	$\frac{29.574 \text{ mL}}{1 \text{ ounce}}$
ounce to g	$\frac{28.35 \text{ g}}{1 \text{ ounce}}$

Convert "from" "to"	Conversion Factor
lb to kg	$\frac{1\ \text{kg}}{2.2046\ \text{lb}}$
mile to km	$\frac{1.6093\ \text{km}}{1\ \text{mile}}$
mile to m	$\frac{1{,}609.3\ \text{m}}{1\ \text{mile}}$
quart to L	$\frac{1\ \text{L}}{1.06\ \text{quarts}}$
gallon to L	$\frac{1\ \text{L}}{0.264\ \text{gallon}}$
mL to cubic cm	$\frac{1\ \text{cm}^3}{1\ \text{mL}}$
L to cubic cm	$\frac{1{,}000\ \text{cm}^3}{1\ \text{L}}$

Metric to English

Convert "from" "to"	Conversion Factor
cm to inch	$\frac{1\ \text{in}}{2.54\ \text{cm}}$
meter to feet	$\frac{3.28\ \text{feet}}{1\ \text{meter}}$
meter to yard	$\frac{1\ \text{yd}}{0.9144\ \text{m}}$
mL to ounce	$\frac{1\ \text{ounce}}{29.574\ \text{mL}}$
g to ounce	$\frac{1\ \text{ounce}}{28.35\ \text{g}}$
kg to lb	$\frac{2.2046\ \text{lb}}{1\ \text{kg}}$
km to mile	$\frac{1\ \text{mile}}{1.6093\ \text{km}}$
m to mile	$\frac{1\ \text{mi}}{1{,}609.3\ \text{m}}$

Convert "from" "to"	Conversion Factor
L to quart	$\frac{1.06 \text{ quarts}}{1 \text{ L}}$
L to gallon	$\frac{0.264 \text{ gallon}}{1 \text{ L}}$
cubic cm to ml	$\frac{1 \text{ mL}}{1 \text{ cm}^3}$
cubic cm to L	$\frac{1 \text{ L}}{1{,}000 \text{ cm}^3}$

The process will be to choose a sequence of conversion factors where each factor cancels out one of the units and this continues until you end up with the desired unit. We shall illustrate this through several examples.

EXAMPLE 1: Convert 2 quarts into Liters.

The process requires us to change quarts to LL. This requires the conversion factor

$$\frac{1 \text{ L}}{1.06 \text{ quarts}}$$

Now multiply by the conversion factors and cancel the units to obtain:

$$2 \text{ quarts} = \frac{2 \cancel{\text{quarts}}}{1} \cdot \frac{1 \text{ L}}{1.06 \cancel{\text{quarts}}} = \frac{2 \text{ L}}{1.06} = 1.89 \text{ L}$$

EXAMPLE 2: Convert 3.5 miles into km.

The process requires us to change miles to km. This requires the conversion factor of

$$\frac{1.6093 \text{ km}}{1 \text{ mile}}$$

Now multiply by the conversion factor and cancel the units to obtain:

$$3.5 \text{ miles} = \frac{3.5 \cancel{\text{miles}}}{1} \cdot \frac{1.6093 \text{ km}}{1 \cancel{\text{mile}}} \approx 5.6 \text{ km}$$

EXAMPLE 3: Convert 450 cubic centimeters (cc) to mL.

The process requires us to change cc's (cubic centimeters) to mL.

This requires the conversion factor, $\dfrac{1\text{ mL}}{1\text{ cm}^3}$

Now multiply by the conversion factor and cancel the units to obtain:

$$450\text{ cc} = \frac{450\ \cancel{\text{cc}}}{1}\cdot\frac{1\text{ mL}}{1\ \cancel{\text{cc}}} = 450\text{ mL}$$

EXERCISES 7.3

Conversions Within the Metric System: Convert the following measurements.

1. 22 m to mm
2. 275 g to kg
3. 15 cm to m
4. 0.15 m to cm
5. 0.034 m to mm
6. 3,200 km to m
7. 68 m to km
8. 22 L to mL
9. 9 mm to cm
10. 0.2 cm to mm

Conversions Within the English System: Convert the following measurements.

11. 2 miles to feet
12. 4.5 yards to feet
13. 15 quarts to gallons
14. 1.25 miles to feet
15. 6 yards to inches
16. 1.5 miles to yards
17. 74 ounces to quarts
18. 2.5 yards to inches

Conversions Between the English and Metric Systems: Convert the following measurements.

19. 22 inches to cm
20. 2 miles to km
21. 38 km to miles
22. 4952 meters to miles
23. 237 feet to meters
24. 47 mL to ounces
25. 5 gallons to Liters
26. 68.9 cm to inches
27. 597 meters to feet
28. 69 fluid ounces to milliliters
29. 190 pounds to kilograms
30. 1843 ounces to kg
31. 197 mm to inches
32. 0.5 liters to ounces
33. 45,000 grams to pounds
34. 75 mph to kph

Applications:

35. One of the largest hailstones ever measured weighed 0.77 kg. How many pounds does that represent?

36. The geostationary weather satellites orbit above the equator at an altitude of 22,375 miles. What is it in km?

37. The National Weather Service often uses a radar frequency whose wavelength is 10 cm to detect precipitation. What is the wavelength in inches?

38. A man is 5 feet 11 inches tall. How many centimeters is this man's height? How many meters is the man's height?

39. If the field diameter of a microscope lens is known to be 4.65 mm, and you can fit 9 copies of an object across the diameter if they are placed end-to-end, how long is the object in mm?

Chapter 7 Practice Test

1. Find the perimeter of a rectangle with a length of 7.6 cm and width of 2.4 cm.
2. Find the area of a square with a side of 7.5 inches.
3. The base of a triangle is 5.8 inches. If the area of the triangle is 13.34 in^2, what is the height of the triangle?
4. A box has dimensions of 5.2 ft by 1.8 ft by 3 ft. Find the volume of the box?
5. What is the area of a rectangle that has length of $3\frac{5}{8}$ cm by $1\frac{2}{5}$ cm?
6. The area of a triangle is $66\frac{1}{2}$ m^2. If the height of the triangle is 8 m, what is the base of the triangle? Leave your answer in fraction form.

Convert the following measurements.

7. 250 g to kg
8. 0.35 m to cm
9. 32 km to m
10. 55 L to mL
11. 0.5 cm to mm
12. 5.4 yards to feet
13. 2 miles to yards
14. 4.8 yards to inches
15. 2.5 miles to km
16. 4952 meters to miles
17. 34.7 cm to inches
18. 1843 ounces to kg
19. 60mph to kph
20. 30 ounces to liters

21. Mary is 5 feet 8 inches tall. What is her height in meters?

22. If the field diameter of a microscope lens is known to be 4.65mm, and you can fit 3 copies of an object across the diameter if they are placed end-to-end, how long is the object in mm?

CHAPTER 8

Negative Numbers – Numbers is Reverse

8.1 INTRODUCTION

In this section we extend our number system to include negative numbers. Negative numbers were first introduced so that business people could keep track of debts. Originally, the mathematical community rejected considering these quantities as numbers. Instead they were thought of as abominations. Eventually, however, their usefulness in being able to quantify important concepts and ideas became obvious, and they were added to the diversity of numbers we have today.

The easiest way to think about negatives numbers is as a reflection of the number line of positive numbers about zero. For example, consider the positive number line:

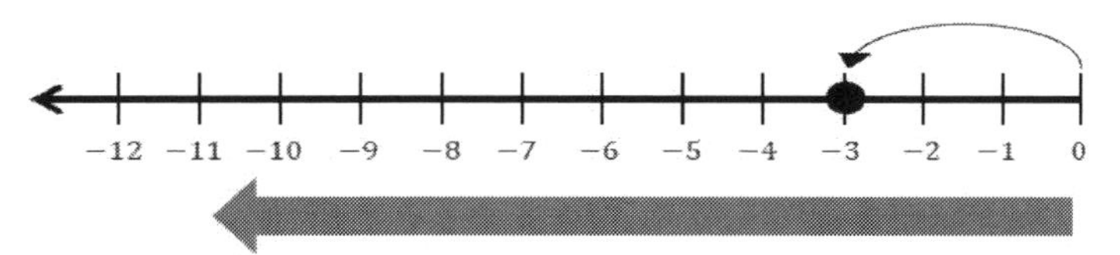

We now have a number line stretching in two directions.

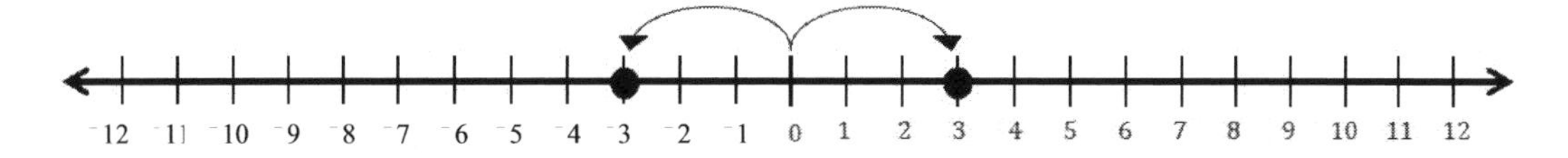

8.2 ORDER AND NEGATIVE NUMBERS

The first thing we need to be able to do is to locate numbers on the number line. To do this we start at zero, and use the fact that positive numbers will always be located to the right of zero and negative numbers to the left.

Thus, positive 4 (+4) is shown as being located fours units to the right of zero below, and negative 4 (¯4) is shown four units to the left of zero.

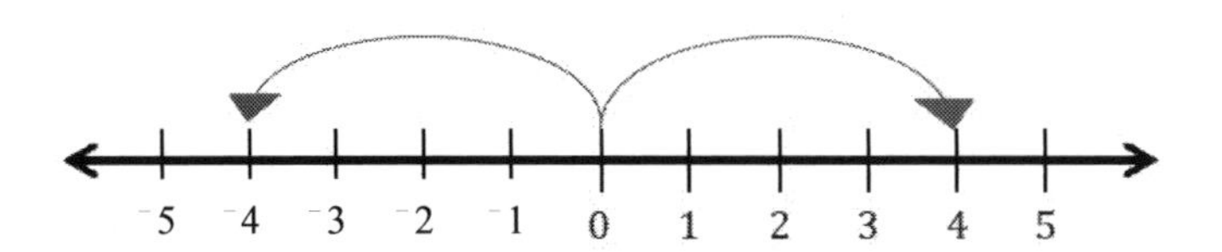

When we considered positive numbers, numbers located to the left of another number were considered less than that number. In this way, negative numbers are considered less than positive numbers. Also, the larger negative numbers are considered less than smaller negative numbers, i.e. $^{-}6 < {^{-}3}$ since $^{-}6$ is to the left of $^{-}3$ on the number line.

Negative numbers can also be thought of as opposite numbers, and every positive number has an opposite negative number on the other side of zero. It is the same distance away from zero, but only on the left side of zero instead of the right.

Consider the following examples where we place the numbers in order using $<$ or $>$.

EXAMPLES: Given the following pairs of numbers, use $<$ or $>$ to order them.

1. $^{-}4$ and $^{-}8$

   $^{-}4 > {^{-}8}$, since $^{-}8$ is to the left of $^{-}4$ when placed on the number line

2. 5 and $^{-}7$

   $5 > {^{-}7}$, since 5 is to the right of $^{-}7$ when placed on the number line

3. $^{-}6$ and 3

   $^{-}6 < 3$, since $^{-}6$ is to the left of 3 when placed on the number line

4. 0 and $^{-}2$

   $0 > {^{-}2}$, since $^{-}2$ is to the left of 0 when placed on the number line

You can think of $<$ and $>$ as arrows that always point to the number that is furthest to the left on the number line.

EXERCISES 8.2

Order and Signed Numbers

Given the following pairs of numbers, use $<$ or $>$ to order them.

1. $^{-}6$ and $^{-}10$
2. $^{-}8$ and $^{-}14$
3. 11 and $^{-}11$
4. 12 and $^{-}3$
5. $^{-}1$ and 1
6. $^{-}9$ and 12
7. 0 and $^{-}5$
8. 0 and $^{-}10$
9. $^{-}2$ and 0
10. $^{-}5$ and 0

8.3 ADDITION AND SUBTRACTION OF NEGATIVE NUMBERS

Adding and subtracting signed numbers can be difficult for students to understand. To help with the concept we will use the number line to guide us.

Adding Signed Numbers

In the addition of signed numbers we always have to remember that adding a positive number will move us to the right on a number line, while adding a negative number will move us to the left on the number line.

Let's consider some examples.

EXAMPLES: Solve the following addition problems.

1. Add: $^{-}4 + 8$

We start with our number line at 0. With the $^{-}4$ we move 4 units to the left. Then adding the 8 we move 8 units to the right leaving us at the location 4 on our number line.

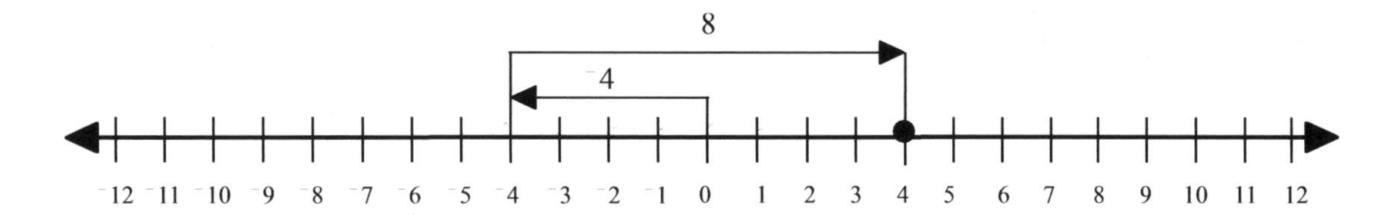

So, $^{-}4 + 8 = 4$

2. Add: $^{-}3 + {}^{-}5$

We start with our number line at 0. With the $^{-}3$ we move 3 units to the left. Then adding the $^{-}5$ means we move 5 units further to the left leaving us at the location $^{-}8$ on our number line.

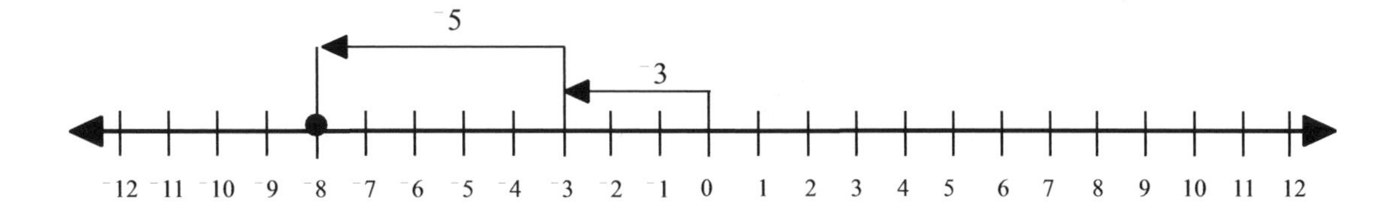

So, $^{-}3 + {}^{-}5 = {}^{-}8$

3. Add: $7 + {}^{-}11$

We start with our number line at 0. With the 7 we move 7 units to the right. Then adding the $^{-}11$

means we now move 11 units to the left leaving us at the location $^{-}4$ on our number line.

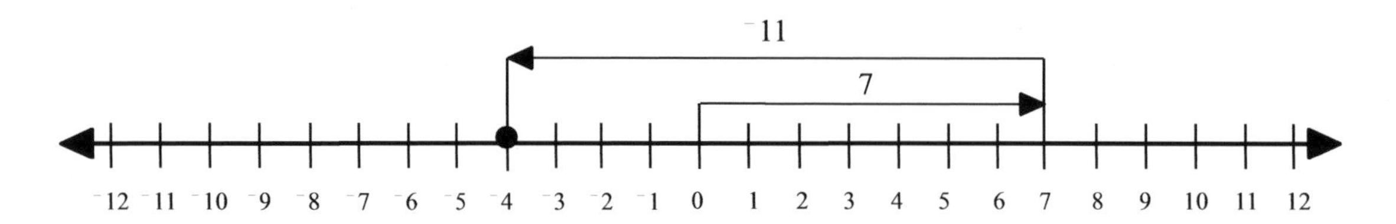

So, $7 + {}^{-}11 = {}^{-}4$

Subtracting Signed Numbers

In the subtraction of signed numbers we always have to remember that subtracting a positive number will move us to the left on a number line, while subtracting a negative number will move us in the opposite direction, or to the right on the number line. It is just the opposite of addition. Addition and subtraction are inverse operations of each other.

We must remember that subtracting a negative is equivalent to adding a positive. Subtracting a debt is the same as adding an asset.

Subtraction is always the more difficult concept than addition. It takes a lot of practice for you to get used to it, and anytime you get confused you should picture the concept using the number line and that should help. The process of double negatives is confusing and not natural.
Let's consider some examples.

EXAMPLES: Solve the following subtraction problems.

1. Subtract: ${}^{-}4 - 8$

We start with our number line at 0. With the $^{-}4$ we move 4 units to the left. Then since we are subtracting 8 we move an additional 8 units to the left from the $^{-}4$ location. Our final location is at $^{-}12$ on the number line.

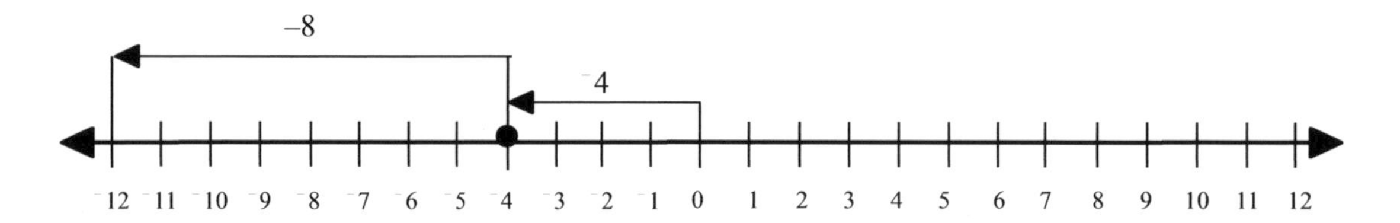

So, ${}^{-}4 - 8 = {}^{-}12$

We want to make a few observations here before we continue. Notice that we chose to put minus 8 (–8) to move 8 units to the left for subtracting 8 positive units on the number line chart above. We can then see that subtracting a positive 8 is equivalent to adding a $^{-}8$. Now subtracting 8 and adding a $^{-}8$ are not the exact same processes, but they are equivalent. This means that we can actually turn all subtraction problems into equivalent addition problems, by simply adding the opposite sign number we are subtracting. If we are subtracting a positive, this is same as adding a negative. Also, as we'll show in the next example, subtracting a negative is equivalent to adding a positive.

Thus, $^{-}4 - 8$ is equivalent to $^{-}4 + {}^{-}8$. That means we can turn all our subtraction problems into equivalent addition problems and solve them as the addition problems above.

To make this more precise, let's look at another few examples.

2. Subtract: $^{-}9 - {}^{-}7$

We start with our number line at 0. With the $^{-}9$ we move 9 units to the left. Then since we are subtracting $^{-}7$ we move in the opposite direction of $^{-}7$, so we move a positive 7 units to the right! Again, subtracting a negative number is equivalent to adding its opposite.

Our final location is at $^{-}2$ on the number line.

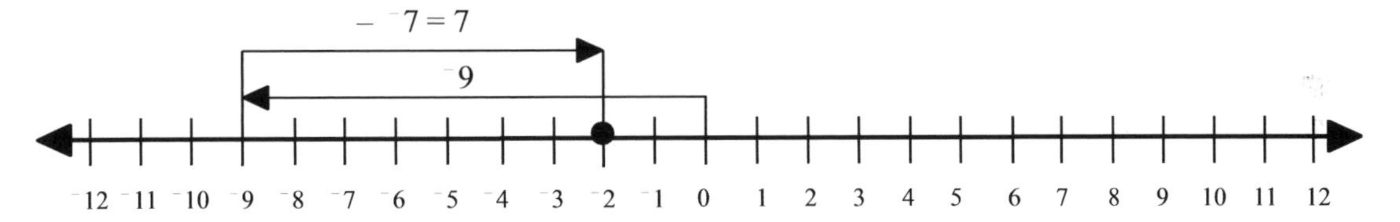

So, $^{-}9 - {}^{-}7 = {}^{-}2$

Again, this is equivalent to the addition problem $^{-}9 + 7 = {}^{-}2$

3. Subtract: $3 - {}^{-}2$

We start with our number line at 0. With the 3 we move 3 units to the right. Then since we are subtracting $^{-}2$ we move in the opposite direction of $^{-}2$, so we move a positive 2 units further to the right! Again, subtracting a negative number is equivalent to adding its opposite.

Our final location is at 5 on the number line.

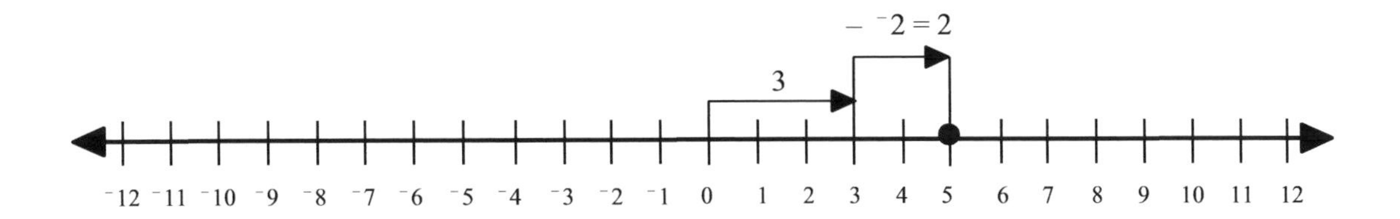

So, $3 - {}^{-}2 = 5$

Again, this is equivalent to the addition problem $3 + 2 = 5$

4. Subtract: $^{-}6 - {}^{-}13$

For this problem we choose to rewrite the subtraction problem into an equivalent addition problem. Thus, $^{-}6 - {}^{-}13$ becomes $^{-}6 + 13$. We can then work the easier addition problem as shown below.

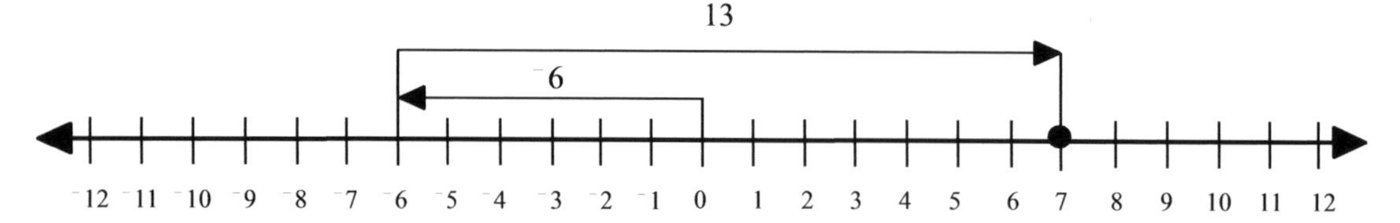

We should point out that we do not always have to draw a number line to do either our addition or subtraction problems. We are doing it here to simply reinforce both the concepts of addition and subtraction, especially when it involves negative numbers.

Ordinarily we'd proceed by changing the subtraction into addition and then solving the addition problem without the number line diagram. If, however, you find it helpful, continue to use it until you no longer have to, or when you get a particularly challenging problem.

Here is the standard algorithm approach.

$$^{-}6 - {}^{-}13 = {}^{-}6 + 13 = 7$$

or

$$^{-}6 - {}^{-}13 = {}^{-}6 + 13 = 13 - 6 = 7$$

5. Subtract: $6 - 11$

In this example we are subtracting two positive numbers, but there is a twist. First we start at 0 and move 6 units to the right. Then we move 11 units to the left. This leaves us at the location of $^{-}5$. Whenever we subtract more than we have, we get a negative result.

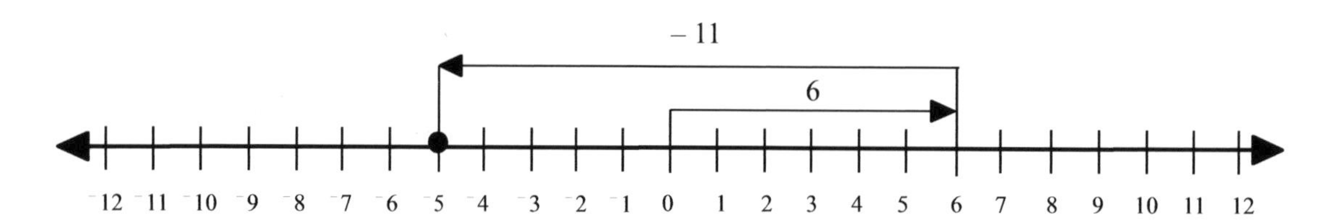

We also could have written this as an addition problem: $6 - 11 = 6 + {}^{-}11 = {}^{-}5$, and obtained the same result.

EXERCISES 8.3

Adding and Subtracting Signed Numbers

Addition

Perform the additions below and justify your response using a number line.

1. $^{-}3 + 6$
2. $^{-}5 + 9$
3. $^{-}5 + {}^{-}4$
4. $^{-}8 + {}^{-}2$
5. $8 + {}^{-}13$
6. $5 + {}^{-}9$

Subtraction

Perform the additions below and justify your response using a number line.

7. $^{-}8 - 2$
8. $^{-}9 - 1$
9. $5 - {}^{-}4$
10. $7 - {}^{-}3$
11. $^{-}11 - {}^{-}6$
12. $^{-}1 - {}^{-}9$

Addition and Subtraction

Perform the additions and subtractions below. You do not have to draw a number line for justification.

13. $^{-}13 - {}^{-}25$
14. $^{-}33 - {}^{-}53$
15. $^{-}23 + {}^{-}35$
16. $^{-}72 + {}^{-}19$
17. $55 - {}^{-}43$
18. $76 - {}^{-}83$
19. $^{-}43 + 25$
20. $^{-}82 + 79$
21. $^{-}85 - {}^{-}115$
22. $^{-}63 - {}^{-}3$
23. $^{-}8 - {}^{-}1\ 7$
24. $64 - {}^{-}64$
25. $36 - {}^{-}36$
26. $16 - 39$
27. $11 - 56$
28. $18 - {}^{-}23$
29. $54 - {}^{-}76$
30. $3.9 - {}^{-}2.6$
31. $^{-}5.75 - {}^{-}3.69$
32. $^{-}34.6 + {}^{-}21.8$
33. $25.39 + {}^{-}32.06$
34. $^{-}21.324 - {}^{-}42.61$
35. $^{-}18.05 - {}^{-}32.3$
36. $\frac{-3}{4} + \frac{1}{4}$
37. $\frac{-4}{5} + \frac{2}{5}$
38. $\frac{-3}{2} - \frac{1}{2}$
39. $\frac{-1}{5} - \frac{2}{5}$
40. $\frac{-1}{2} + \frac{-2}{3}$
41. $\frac{-3}{5} + \frac{-2}{7}$
42. $\frac{-1}{3} + \frac{-4}{5}$
43. $\frac{-1}{5} + \frac{-2}{3}$
44. $\frac{-1}{2} - \frac{-2}{3}$
45. $\frac{-3}{5} - \frac{-2}{7}$
46. $\frac{-3}{4} - \frac{-5}{6}$
47. $\frac{-2}{7} - \frac{-5}{8}$

8.4 MULTIPLICATION AND DIVISION OF NEGATIVE NUMBERS

Multiplication and division of signed numbers is a very confusing concept. In fact one of the greatest mathematicians of all time, Leonard Euler, was not able to prove one of the basic facts when multiplying two negative numbers. He, however, used the fact without proof. We should point out that using the axioms of our Hindu-Arabic number system, we can prove all the results we use. We will omit the formal proof though, since it will confuse some of you. We will, however, try and conceptually justify all the ideas we present below. Let's begin with multiplication.

Multiplication

Let's start with multiplying a positive number times a negative number, for example:

$$3 \times {}^{-}4$$

Now we don't know what to do with a positive times a negative, but we do have rules and definitions on how to proceed. There is one rule, or axiom, and one definition we need to move forward. The firs rule is called the distributive property, which tells us how to multiply a number times a sum of two numbers.

For example consider: $3 \times (5+2)$ The distributive property tells us we get the same result whether we add the 5 and the 2 and then multiply, or if we distribute the 3 to the 5 and then the 2 and add the two products together, i.e.

$$3\times(5+2) = 3 \times 5 + 3 \times 2 = 6 + 15 = 21$$
$$= 3\times7 = 21$$

That is the distributive property.

The other thing we use is a definition which says that every number has something defined to be its additive inverse. An additive inverse is a number that when added to your number gives zero as a result. For example the additive inverse of 4 is ${}^{-}4$, since $4 + {}^{-}4 = 0$. We just did addition in the last section.

Using these two aspects of our number system we can now answer our original question; What is $3 \times {}^{-}4$?

First, let's start with another fact that any number times zero is equal to zero. That is how we define it, or more specifically:

$$3 \times 0 = 0$$

Next, we use our additive inverse to replace zero with what it's equivalent to, namely $4 + {}^{-}4 = 0$.

$$3 \times (4 + {}^{-}4) = 0$$

We now apply the distributive property to obtain:

$$3 \times 4 + 3 \times {}^{-}4 = 0$$

We can do the first product, since that is just two positive numbers.

$$12 + 3 \times {}^{-}4 = 0$$

We now have that 12 added to a number must equal zero. By definition that number must be its additive inverse or ${}^{-}12$!

That means we can now answer the question of what is $3 \times {}^{-}4$ it is ${}^{-}12$, or

$$3 \times {}^{-}4 = {}^{-}12$$

So a positive number times a negative number is a negative number. Now we could do a similar justification to show that a negative times a positive is also negative.

The next question we ask is what is a negative times a negative number, for example:

$${}^{-}6 \times {}^{-}9$$

Here we follow the same approach and start with

$${}^{-}6 \times (9 + {}^{-}9) = 0$$

Distribute the ${}^{-}6$ to obtain

$${}^{-}6 \times 9 + {}^{-}6 \times {}^{-}9 = 0$$

Since we now know that ${}^{-}6 \times 9 = {}^{-}54$ we can replace that value to obtain

$${}^{-}54 + {}^{-}6 \times {}^{-}9 = 0$$

This is our additive inverse statement again, so that we are looking for the additive inverse of ${}^{-}54$, which is 54. This tells us that

$${}^{-}6 \times {}^{-}9 = 54$$

Thus, a negative times a negative is positive!

Now, this is not a formal proof, but we hope it provides some justification for this sometimes confusing fact in mathematics. It is not the consequence of any analogy we may try and conjure up, it is instead a fact based upon the rules of arithmetic that arise from our basic axioms and definitions.

With these two rules we can now proceed in a very simple way with the multiplication of signed numbers.

Rule 1: A negative times a positive or a positive times a negative is equal to a negative of the product of the two numbers without their signs.

Rule 2: A negative times a negative is equal to a positive of the product of the two numbers without their negative signs.

EXAMPLES: Calculate the products below.

1. Multiply: ${}^{-}6 \times 3$

Since this is a product of a negative times a positive we use Rule 1 to compute this as

$$^{-}(6 \times 3) = {}^{-}18$$

2. Multiply: $5 \times {}^{-}13$

Since this is a product of a negative times a positive we use Rule 1 to compute this as

$$^{-}(5 \times 13) = {}^{-}65$$

3. Multiply: ${}^{-}5 \times {}^{-}7$

Since this is a product of a negative times a negative we use Rule 2 to compute this as

$${}^{-}5 \times {}^{-}7 = 5 \times 7 = 35$$

The same holds true for decimals and fractions.

4. Multiply: $^{-}\frac{3}{5} \times {}^{-}\frac{2}{7}$

Since this is a product of a negative times a negative we use Rule 2 to compute this as

$$\frac{3}{5} \times \frac{2}{7} = \frac{6}{35}$$

5. Multiply: $2.5 \times {}^{-}1.3$

Since this is a product of a negative times a positive we use Rule 1 to compute this as

$$^{-}(2.5 \times 1.3) = {}^{-}3.25$$

We now move on to division.

Division

We will use the earlier fact that we developed showing us that division is equivalent to multiplying by the reciprocal of what we are dividing by. Since division can be transformed into an equivalent multiplication sentence, we argue that it makes sense that division should follow the same rules as multiplication for signed numbers.

Thus, we then have the following rules for division of signed numbers.

Rule 3: A negative divided by a positive or a positive divided by a negative is equal to the negative of the quotient of the two numbers without the negative sign.

Rule 4: A negative divided by a negative is equal to a positive of the quotient of the two numbers without their negative signs.

With these two rules we can proceed with some examples.

EXAMPLES: Calculate the quotients below.

1. Divide: $^-6 \div 3$

Since this is a quotient of a negative divided by a positive we use Rule 3 to compute this as

$$^-(6 \div 3) = {}^-2$$

2. Divide: $^-18 \div {}^-3$

Since this is a quotient of a negative divided by a negative we use Rule 4 to compute this as

$$18 \div 3 = 6$$

3. Divide: $24 \div {}^-3$

Since this is a quotient of a positive divided by a negative we use Rule 3 to compute this as

$$^-(24 \div 3) = {}^-8$$

4. Divide: $^-16 \div 12$

Since this is a quotient of a negative divided by a positive we use Rule 3 to compute this as

$$^-(16 \div 12) = {}^-\frac{16}{12} = \frac{4\times 4}{3\times 4} = {}^-\frac{4}{3}$$

5. Divide: $^-9.3 \div {}^-3.1$

Since this is a quotient of a negative divided by a negative we use Rule 4 to compute this as

$$9.3 \div 3.1 = 3.0$$

6. Divide: $\frac{3}{5} \div {}^-\frac{2}{7}$

Since this is a quotient of a positive divided by a negative we use Rule 3 to compute this as

$$^-\left(\frac{3}{5} \div \frac{7}{2}\right) = {}^-\frac{3\times 7}{5\times 2} = {}^-\frac{21}{10}$$

EXERCISES 8.4

Multiplying and Dividing Signed Numbers

Perform the indicated operations below.

1. $^{-}7 \times 5$
2. $^{-}4 \times 9$
3. $^{-}6 \div 2$
4. $^{-}21 \div 7$
5. $5 \times {}^{-}9$
6. $12 \times {}^{-}7$
7. $^{-}54 \div {}^{-}27$
8. $^{-}48 \div {}^{-}8$
9. $^{-}3 \times {}^{-}32$
10. $^{-}6 \times {}^{-}13$
11. $50 \div {}^{-}15$
12. $35 \div {}^{-}21$
13. $14 \times {}^{-}8$
14. $23 \times {}^{-}2$
15. $^{-}15 \div 9$
16. $^{-}36 \div 27$
17. $^{-}18.03 \div {}^{-}8.45$
18. $^{-}26.118 \div {}^{-}3.82$
19. $9.2 \times {}^{-}3.5$
20. $11.3 \times {}^{-}8.7$
21. $\frac{7}{12} \div {}^{-}\frac{5}{12}$
22. $\frac{9}{13} \div {}^{-}\frac{3}{26}$
23. $^{-}\frac{5}{22} \div {}^{-}\frac{10}{11}$
24. $^{-}\frac{7}{5} \times {}^{-}\frac{15}{14}$
25. $^{-}4 \times {}^{-}\frac{5}{8}$
26. $^{-}12 \times {}^{-}\frac{7}{9}$

8.5 ABSOLUTE VALUE

Sometimes we may need to consider the size of a debt or asset. With this type of problem we are not concerned with its order or precise location on the number line, but only how far it is away from zero. Thus, the sign of the number is not important, only its magnitude, i.e. how far it is away from zero.

The distance finding operator is given the name of the absolute value and it is calculated as follows. If the number is positive, we do nothing and just return the positive number. If, however, the number is negative, we return the opposite (recall, the opposite of a number means change its sign) of the negative number. That means we change (drop) the sign of the number and return the resulting positive value (magnitude) of that negative number.

We should point out that distance (a measure of how far away something is) is always positive. We would never say that we are negative four feet away from an object. We would simply say that we are four feet away from it. Thus, for example the numbers 4 and $^{-}4$ are both four units away from zero as shown in the figure below. The number $^{-}4$ just so happens to be four units to the left of 0 and 4 is simply four units to the right of 0. The sign just tells us its direction (location) relative to zero.

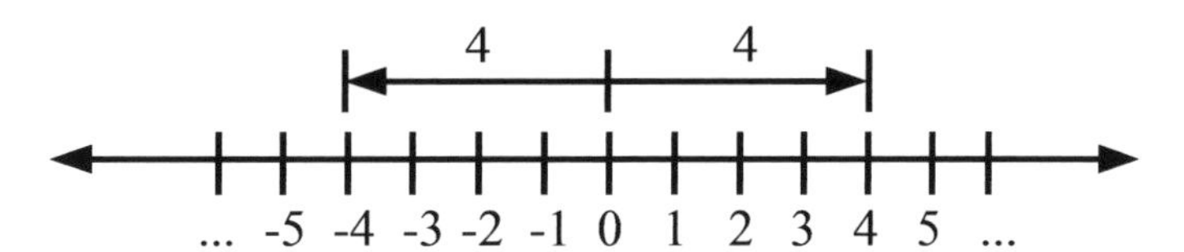

We can represent the absolute value operator as two vertical lines surrounding the number, |number|. Thus, when we see $|{}^{-}2|$, we recognize that we are finding the distance the number $^{-}2$ is from the origin and, since the number is negative, we change the sign and the answer is 2.

EXAMPLE 1: Evaluate $|3|$.

Solution: $|3| = 3$, since 3 is not negative we leave the sign the same.

EXAMPLE 2: Evaluate $|{}^{-}13.9|$.

Solution: $|{}^{-}13.9| = 13.9$, since $^{-}13.9$ is negative we change the sign to positive.

EXAMPLE 3: Evaluate $|0|$.

Solution: $|0| = 0$, since 0 is neither negative nor positive we leave it as it is.

EXAMPLE 4: Evaluate $|{}^{-}\pi|$.

Solution: $|{}^{-}\pi| = \pi$, since $^{-}\pi$ is negative we change the sign to positive.

EXERCISES 8.5

Absolute Value

Evaluate each expression

1. $|2|$
2. $|-7|$
3. $-|-5|$
4. $|-\pi|$
5. $-|17|$
6. $\left|\frac{2}{3}\right|$
7. $\left|-\frac{5}{2}\right|$
8. $|-45.7|$
9. $|29.3|$
10. $-\left|-\frac{3}{7}\right|$
11. $-|-12|$
12. $\left|\frac{1}{2}\right|$
13. $\left|\frac{15}{16}\right|$
14. $|-\sqrt{2}|$
15. $|3.14|$
16. $-|-67.9|$
17. $|-5|$
18. $|9|$
19. $|1.23|$
20. $|-56|$
21. $|-7|$

22. $|-4.7|$

23. $\left|\frac{3}{8}\right|$

24. $\left|-\frac{1}{3}\right|$

25. $|0.45|$

26. $|-8.4|$

Chapter 8: Practice Test

Given the following pairs of numbers, use < or > to order them.

1. $^{-}17$ and $^{-}12$
2. 6 and $^{-}9$
3. 114 and $^{-}15$

Perform the additions below and justify your response using a number line.

4. $^{-}8 + 11$
5. $^{-}3 + ^{-}7$
6. $^{-}8 - 6$
7. $^{-}9 - 3$
8. $8 - ^{-}4$

Perform the indicated operations below.

9. $^{-}13 + ^{-}25$
10. $^{-}33 - ^{-}33$
11. $^{-}7 \times 5$
12. $^{-}6 \div ^{-}2$
13. $^{-}7 \div 21$
14. $^{-}3 - ^{-}35$
15. $32 \times ^{-}4$
16. $^{-}54 \div ^{-}9$
17. $^{-}31 + 31$
18. $^{-}3 \times ^{-}14$
19. $4 \times ^{-}82$
20. $^{-}13 + ^{-}25$
21. $^{-}8.3 \times ^{-}3.5$
22. $^{-}6.39 \div ^{-}3.13$
23. $1.3 \times ^{-}1.7$
24. $\frac{7}{12} + \frac{^{-}5}{12}$
25. $\frac{^{-}5}{22} \div \frac{^{-}15}{44}$

CHAPTER 9

Variables – Numbers in Disguise

This is where arithmetic ends and some basic algebra begins. It is with the concept of the variable that the true power of arithmetic becomes unleashed through the algebra. In this chapter we introduce the variable concept.

9.1 WHAT ARE VARIABLES

The word vary, obviously describes something that changes. In algebra, this **something** is a number. Now, instead of the fixed numbers of arithmetic, we want to talk about numbers more abstractly. We don't want to consider the exact number two, three, one-half, etc. Instead we wish to talk about numbers more generally.

For example, if you wanted to discuss the dimensions of a room, you would say the room has a certain length and width, and that together the length and width provide the dimensions of the room. You could then compare rooms of different sizes, depending upon what their respective lengths and widths were. In talking about the various lengths and widths, you are actually using the fundamental concept of algebra called a variable. Since, we may not know the precise length or width of the room ahead of time, we give these unknown values names, such as length and width. These values will vary depending upon the room we are talking about. Now, rather than using the whole name to designate length and width, in mathematics we typically use the abbreviations ℓ for length, and w for width to represent these varying numerical quantities.

Think about other numerical quantities that you work with, without even knowing their values – someone's height or weight, the temperature during the day, the price of a shirt at various stores, the cost of a particular item of food at different supermarkets, the value of a particular stock on any given day, etc. You are constantly using the concept of a variable without even thinking about it. Any time you think about something that can be quantified numerically, you are taking into consideration that the value may vary. This is the concept of a variable. Any numerical quantity that can change can be thought of as a variable.

In this chapter, we wish to build upon this intuitive concept and extend it so that it becomes much more useful as a predictive tool. Wouldn't it be nice to see a trend in how a price was changing, and be able to react or do something about it?

Doctors know that it's helpful to measure and compare various chemical levels in peoples' bodies to ascertain what levels of which chemicals are good, and which are bad. Without the concept of variable, our health would deteriorate! We wouldn't be able to keep track of changes in our bodies.

What makes algebra very useful, but at the same time more confusing, is when we change the focus from what the variable actually represents – time, distance, money, temperature, etc., and instead focus on its properties as a variable. Thus, we strip away the **physical** meaning and are then able to focus on how this now abstract concept of a variable can be manipulated to obtain new and interesting results, which can then be re-applied and re-interpreted in terms of its original **physical** meaning at a later date should we so desire. This is both a curse and a blessing.

The convention has been to choose letters near the end of the alphabet, such as x, y, z, to represent variables that do not have a concrete meaning assigned to them. Variables that do have a specific meaning are sometimes assigned the value of the first letter of the word used to describe them – ℓ for length, w for width, t for time, A for amount or area, V for volume, C for cost or circumference, P for profit or pressure, etc.

In this chapter we will investigate the arithmetic properties of these unassigned/abstract variables. To simplify the explanation, we will only consider problems involving a single variable. In the later part of the book we will begin to look at problems with two or more variables, but for now it is best to develop this new language using only a single variable.

Definition: A **Variable** is a letter used to represent an unspecified number.

EXAMPLES: x, y, z, a, b, c, N, W, etc.

As with any number, we can add, subtract, multiply and divide variables. Thus, if we have a variable, x, then we can multiply x by a number, say three, and obtain a new number that can be written as either $3\cdot x$, or $3\times x$ (which is rarely used, since the multiplication symbol looks too much like the variable "x."), or $3 * x$, or as we will prefer to write it 3x. You should make the distinction that this is really 3 times an unknown quantity x, and not simply a new unknown 3x quantity. The 3 and the x are separable.

Furthermore, addition, subtraction, and division can be represented as: $x + 2$ or $2+x$, $x-2$ or $2-x$, $x \div 2$ or $x/2$, or $\frac{x}{2}$, respectively.

To help our discussion move forward, we need to begin to start attaching names to these new quantities we begin to develop. By combining variables with arithmetic, we create what we call terms and expressions.

Definition: A **term** is either a single fixed number, also called a constant, or a constant either multiplying or dividing a variable.

EXAMPLES: $^{-}13$, $\frac{7}{12}$, $2x$, or $\frac{3}{5}x$

Definition: An **expression** is simply a sum of terms.

EXAMPLES: $2+3x$,
$4-3x$ which can be rewritten as $4 + (-3x)$, or $4 + {}^{-}3x$,
$2x-5+x$, which can be rewritten as $2x + {}^{-}5 + x$,

$\frac{1}{3}x+4-3x$ which can be rewritten as, $\frac{1}{3}x+4+(^{-}3x)$

$\frac{x}{2}+7-3+4x$ which can be rewritten as $\frac{x}{2}+7+(^{-}3)+4x$

9.2 EVALUATING EXPRESSIONS

We will often need to find the fixed numerical value of an expression, given a fixed value for the variable in the expression. To do so, we simply replace the variable with the specified number and perform the arithmetic calculations on the fixed numbers.

EXAMPLE 1: Evaluate, $2x-1$ for $x=3,\ -2,\ 7$

Solution: $x=3$, $2(3)-1=6-1=5$

Solution: $x=-2$, $2(-2)-1=-4-1=-5$

Solution: $x=7$, $2(7)-1=14-1=13$

EXAMPLE 2: Evaluate, $5-3x$ for $x=1,\ -4,\ 0$

Solution: $x=1$, $5-3(1)=5-3=2$

Solution: $x=-4$, $5-3(-4)=5-(-12)=5+12=17$

Solution: $x=0$, $5-3(0)=5-0=5$

EXAMPLE 3: Evaluate, $2x-3$ for $x=1,\ \frac{1}{2},\ -\frac{3}{2}$

Solution: $x=1$, $2(1)-3=2-3=-1$

Solution: $x=\frac{1}{2}$, $2\left(\frac{1}{2}\right)-3=\frac{2}{2}-3=1-3=-2$

Solution: $x=-\frac{3}{2}$, $2\left(-\frac{3}{2}\right)-3=\frac{-6}{2}-3=-3-3=-6$

EXERCISES 9.2

Evaluating Expressions

Evaluate each using the value given.

Basic Problems

1. $2x-3$; use $x=-1$
2. $1-4a$; use $a=7$
3. $-3-12n$; use $n=-3$
4. $5x-3+2x-4$; use $x=4$
5. $(4a+9)2-7$; use $a=2$
6. $-2(2x-3)$; use $x=-7$
7. $4-(1-2b)$; use $b=-5$
8. $(4-7x)$; use $x=0$
9. $5-6a$; use $a=-6$
10. $8-3m$; use $m=\frac{2}{3}$
11. $7x+\frac{3}{5}$; use $x=-\frac{2}{7}$

Intermediate and Advanced Problems

12. $3x-7$; use $x=-\frac{4}{3}$
13. $2-5r$; use $r=\frac{3}{5}$
14. $4+8p$; use $p=-\frac{1}{4}$
15. $6y+12$; use $y=\frac{1}{2}$
16. $10w-8$; use $w=-\frac{1}{5}$
17. $-2-(4+3n)-7$; use $n=8$
18. $4(4x-6)-16x+24$; use $x=3$
19. $4n-6+4(7n-3)$; use $n=4$
20. $-(p+3)+(12-3p)-32$; use $p=-6$
21. $(3x-2)-(2-5x)$; use $x=0$
22. $16+c-3c+(1-2c)$; use $c=-1$
23. $(b-3)3-2b$; use $b=-5$
24. $-\frac{1}{5}(10x-15)+3x$; $x=5$
25. $\frac{2}{3}(-12x-18)-19$; use $x=-\frac{1}{2}$

Chapter 9: Practice Test

Evaluate each of the following expression using the given values.

1. $2x+5$, use $x=3$
2. $5-b$; use $b=-2$
3. $8-3m$; use $m=\frac{4}{3}$
4. $6y+12$; use $y=\frac{1}{3}$
5. $5k-7$; use $k=2.3$
6. $2n-5$; use $n=3$

CHAPTER 10

Ratios, Rates and Proportions

In this chapter we introduce three related ideas. That of a ratio, a rate and a proportion. These important ideas help us connect the concept of fractions to division, and to expand upon the meaning of a fraction.

10.1 WHAT IS A RATIO

The dictionary definition of ratio is: "the quantitative relation between two amounts showing the number of times one value contains or is contained within the other." For example, the ratio of 35 to 5, asks how many fives are in thirty-five, or how many copies of 5 are contained in 35?

Mathematically we are asking the question $? \times 5 = 35$, which is equivalent to to the division statement $35 \div 5 = ?$. This connects the idea of a ratio to the process of division. Thus, more simply, a ratio can be viewed as dividing two numbers.

16 to 2 is equivalent to $16 \div 2$

34 to 17 is equivalent to $34 \div 17$

We now connect the idea of division, which is a verb or an action, to that of a fraction, which is a noun or a name. Most likely you have been doing this all along in your math courses, but never gave it any thought, because you were taught to just do it. Now, however, we want to examine it more clearly, and hopefully you will understand that they are actually two different concepts that can be seen to give equivalent results.

Division of whole numbers is defined as follows:

> The **division of a whole number m by a whole number n > 0**, written as $\mathbf{m \div n}$, is the whole number k that satisfies $m = k \times n$. Where the number k is called the **quotient**, n the **divisor**, and m the **dividend**.

For example, $12 \div 4 = 3$ is equivalent to $12 = 3 \times 4$, or $28 \div 4 = 7$ is equivalent to $21 = 7 \times 4$.

This definition can be extended to include cases where k in not a whole number.

> The **division of a whole number m by a whole number n > 0**, written as $\mathbf{m \div n}$, is by definition the length of one part when a segment of length m is divided into n equal parts.

This new definition requires the introduction of unit fractions, from Chapter 3, as the n equal parts. This idea, coupled with the commutative property of multiplication, then gives a similar multiplication statement, $m = n \times k$, but now k is a unit fraction.

For example, $5 \div 3$ is equivalent to the multiplication statement $5 \times \frac{1}{3} = \frac{5}{3}$, which is essentially the definition of a fraction. This means that $5 \div 3$ is equivalent to the fraction $\frac{5}{3}$, or from now on we can replace one with the other because $5 \div 3 = \frac{5}{3}$.

This means that any ratio, with is one number divided by another, can also be thought of as a fraction.

Using this concept of ratios we can compare the sizes of groups of the same objects.

EXAMPLE 1: Compare the two groups using ratios. 10 people to 7 people.

$$\frac{10 \text{ people}}{7 \text{ people}} = \frac{10 \text{ \sout{people}}}{7 \text{ \sout{people}}} = \frac{10}{7}$$

The units of people cancel, so this is equivalent to saying that the group of 10 people to 7 people is equivalent to writing, $\frac{10}{7}$.

Stated another way, the ratio of 10 people to 7 people is equivalent to the fraction $\frac{10}{7}$. Thus for every 7 people in group 2 there are 10 people in group 1.

EXAMPLE 2: Compare the two groups using ratios. 24 cars to 6 cars.

$$\frac{24 \text{ cars}}{6 \text{ cars}} = \frac{24 \text{ \sout{cars}}}{6 \text{ \sout{cars}}} = \frac{24}{6} = 4$$

The units of cars cancel, so this is equivalent to saying that the group of 24 cars to 6 cars is equivalent to writing, $\frac{24}{6} = 4$. This means that the group with 24 cars is 4 times as large as the group with 6 cars.

Stated another way, the ratio of 24 cars to 6 cars is equivalent to the fraction $\frac{24}{6} = \frac{4}{1} = 4$. Thus for every 6 cars in group in group 2 there are 24 cars in group one, or equivalently, for every 1 car in group 2 there are 4 cars in group 1.

Notice that in the two examples above the final number has no units, it is just a number. All ratios are without units. They simply mean we are comparing the relative size of two groups of the same types of objects.

Another common type of ratio that we introduced in Chapter 6 is the percentage. Percentages are ratios out of a fixed number, 100. As we recall, a percentage can always be written as a fraction over 100, e.g.; $20\% = \frac{20}{100}$, $7\% = \frac{7}{100}$, $93\% = \frac{93}{100}$, etc. Percentages also compare the same objects, only they compare it to a fixed number of 100 objects. We can extend what we do for ratios to include percentages.

EXAMPLE 3: Compare the two groups using ratios. 10 people to 7 people. Then rewrite as a decimal rounded to the nearest hundredths place, then as a fraction over 100, and finally as a percent.

$$\frac{10\text{ people}}{7\text{ people}} = \frac{10\ \cancel{\text{people}}}{7\ \cancel{\text{people}}} = \frac{10}{7} = 1.42857\ldots \approx 1.43 = \frac{143}{100} = 143\%$$

EXAMPLE 4: Compare the two groups using ratios. 8 boats to 15 boats. Then rewrite as a decimal rounded to the nearest hundredths place, then as a fraction over 100, and finally as a percent.

$$\frac{8\text{ boats}}{15\text{ boats}} = \frac{8\ \cancel{\text{boats}}}{15\ \cancel{\text{boats}}} = \frac{8}{15} = 0.5333\ldots \approx 0.53 = \frac{53}{100} = 53\%$$

EXERCISES 10.1

Compare the number of objects in the second group to the number in the first, and then explain what the resulting ratio means when written as a fraction. Also rewrite as a decimal approximated to the hundredths place, then as a fraction over a 100, and finally as a percentage.

1. $32 to $25
2. 36 cases of books to 12 cases of books.
3. 16 bottles to 19 bottles
4. 45 meters to 10 meters
5. 175 cm to 25 cm
6. 28 tickets to 56 tickets
7. 120 square feet to 10 square feet.
8. 4 boats to 3 boats

10.2 WHAT IS A RATE

A rate is simply a ratio where the two quantities we are comparing have different units. In particular, consider the common rate of speed. This is a comparison of a distance (for example miles) to time (or hours). This is called the speed which is characterized by the rate of miles per hour.

In this case the rate, which is the speed, is characterized by a distance traveled (or a change in distance), over an elapsed time (or a change in time). A rate is simply a measure of how the quantity in the numerator (top) changes as the quantity in the denominator (bottom) changes. Now, we frequently think of a rate as something that changes with time, such as:

$$\text{Rate} = \frac{\text{Change in Miles}}{\text{Change in Hours}}$$

$$\text{Rate} = \frac{\text{Change in volume}}{\text{Change in time}}$$

$$\text{Rate} = \frac{\text{Number of heartbeats}}{\text{Number of minutes}}$$

$$\text{Rate} = \frac{\text{Change in the Earth's temperature}}{\text{Change in the number of years}}$$

In all these rates we are considering how the quantity on top (numerator) changes with a changing quantity of time on the bottom (denominator).

However, we can go beyond the "time" interpretation and consider other possible rates-of-change. For example, consider the following rates:

$$\text{Rate} = \frac{\text{Profit from an item}}{\text{Cost of the item}}$$, or, how the profit changes with a changing cost.

$$\text{Rate} = \frac{\text{Temperature}}{\text{Altitude}}$$, or, how the temperature changes with a changing altitude.

Thus, we can think of a rate more generally, as how changing one quantity, changes another quantity.

This is an extremely important concept in both mathematics as well as the sciences. With this in mind, we consider some basic examples that involve finding rates.

We should, however, point out that we are actually finding an **average rate** in all the quantities above. For example, we know that it is possible to drive from location A to location B, and change your speed as you travel along the way. However, if we only consider the total distance traveled divided by the total time to travel it, this will only give us the average speed in which we traveled. It says nothing about how our speed might have changed along the way. This is also true for rates not involving time. Thus, in this section we are only considering average rates and assuming that our speed never changes. We realize that the last assumption is not realistic, but it is necessary at this level of mathematics.

EXAMPLE 1: A plane flies 1,500 miles in 3 hours. What is its average rate of change or speed?

$$\text{rate} = \frac{1500\,\text{miles}}{3\,\text{hours}} = 500\frac{\text{miles}}{\text{hours}} = 500\,\text{mph}$$

This means that on average, the plane flew at 500 mph.

EXAMPLE 2: As we climb 2,000 feet in altitude, the temperature goes down by 35 ^{O}F. What is its average rate-of-change?

$$\text{rate} = \frac{-35^{O}\text{F}}{2{,}000\text{ feet}} = -0.0175\frac{^{O}\text{F}}{\text{foot}}$$

This means that on average, the temperature decreased by 0.0175 ^{O}F per foot of increased altitude.

EXAMPLE 3: In a glycolysis reaction that takes 3.25 hours, the volume of CO_2 produced is 3.5 mL. What is the rate of CO_2 production?

Note: Since we are looking for a rate of production, that implies that we are looking for how much "stuff" is produced over time. Therefore, the rate should be expressed in units of $\frac{\text{volume}}{\text{time}}$.

$$\text{rate} = \frac{3.5\,\text{mL}}{3.25\,\text{hours}} \approx 1.077\frac{\text{mL}}{\text{hour}}$$

This means that the CO_2 production was at an average rate of about 1.077 mL per hour.

EXERCISES 10.2

Given the following, find the rates. Be sure to include the appropriate units.

1. A car travels 345 miles in 6 hours, What is its average rate-of-change in miles per hour?
2. A car travels 745 miles in 12 hours, What is its average rate-of-change in miles per hour?
3. As we climb 3,000 feet in altitude, the temperature goes down by 50 ^{O}F. What is its average rate-of-change of ^{O}F per foot?
4. As we climb 15,000 feet in altitude, the temperature goes down by 75 ^{O}F. What is its average rate-of-change of ^{O}F per foot?
5. On an electrocardiogram, you observe that a patient's normal heartbeat occurs every 0.84 seconds. What is the patient's heart rate in heartbeats per minute?
6. On an electrocardiogram, you observe that a patient's normal heartbeat occurs every 0.75 seconds. What is the patient's heart rate in heartbeats per minute?
7. In a glycolysis reaction that takes 2 hours, the volume of CO_2 produced is 3.25 mL. What is the rate of CO_2 production?
8. In a glycolysis reaction that takes 2.5 hours, the volume of CO_2 produced is 4.75 mL. What is the rate of CO_2 production?
9. In a glycolysis reaction that takes 90 minutes, the volume of CO_2 produced is 2.75 mL. What is the rate of CO_2 production?
10. A competitive high school swimmer takes 52 s to swim 100 yards. What is his rate in yards per second? What is his rate in feet per minute?
11. The *Alvin*, a submersible research vessel, can descend into the ocean to a depth of approximately 4500 m in just over 5 hr. Determine its average rate of submersion in meters per hour. What is

its average rate in kilometers per second?
12. A competitive college runner ran a 5K (5 km) race in 15 min, 25 s. What was her pace in km per min? Meters per second?
13. A compressor will compress 20 liters of air in 15 minutes. What is the rate of compression in liters per hour?
14. A weather balloon rises 275 feet in 15 seconds. How fast is it rising in feet per minute?
15. A parachute descends 255 meters in 8 minute. At what rate, in meters per minute is the parachute descending?
16. In taking a persons pulse you count 15 heartbeats in 10 seconds. What is the heart rate in beats per minute?
17. In taking a persons pulse you count 8 beats in 5 seconds. What is the heart rate in beats per minute?
18. A plane descends 25 miles in 5 minutes. What is the rate is it descending in miles per hour?
19. A runner can run 320 feet in 10 seconds. At what rate in miles per hour are they running?
20. A bird can fly 250 meters in 20 seconds. How fast are they traveling in meters per hour?
21. You ski down a hill and in 15 seconds you travel 110 meters. How fast are you traveling in meters per minute?
22. The potassium permanganate dye in a gel is diffusing at 18 mm every 2 minutes. What is the rate of diffusion in centimeters per hour (cm/hr)?
23. The dye in a gel is diffusing at 12 mm every 3 minutes. What is the rate of diffusion in centimeters per hour (cm/hr)?

10.3 WHAT IS A PROPORTION

The term proportion can either be viewed as a noun or as a verb. For our purposes we shall view it as a verb, meaning there is an action to be performed. More specifically, a proportion is when we set two ratios or rates equal to each other.

For example:

$$\frac{16\text{ miles}}{3\text{ hours}} = \frac{48\text{ miles}}{9\text{ hours}}, \quad \frac{25}{100} = \frac{1}{4}, \quad \frac{\$220}{1\text{ week}} = \frac{\$550}{2.5\text{ weeks}}, \quad \frac{33}{70} = \frac{5}{7}$$

Now, just because we set two ratios or rates equal to each other, does not imply that they are indeed equal. The last proportion in the example above does not have equal ratios. Thus, the first step we must perform is a verification process, where we determine if the two ratios or rates are equal to each other.

This then leads to the more important application of proportions, which is solving a proportion when part of one of the ratios or rates is unknown, and we wish to determine the value of that unknown part which will make the proportion true, or verified.

Proportions either require that we verify that the proportion is true, or that we find a value for an unknown in the proportion that makes it a true statement. The later operation is called solving a proportion. In the next two sections we see how to first verify and then solve proportions.

10.4 VERIFYING PROPORTIONS

When working with a proportion involving two ratios, the verification process is already familiar to us. This is essentially verifying that we have equivalent fractions. The process to verify if two fractions are equivalent, is the cross-multiplication algorithm. The cross-multiplication algorithm states that two fractions (ratios) are equal if, when we cross-multiply (the denominator of one fraction times the numerator of the other fraction), we get equal numbers on both sides of the equal sign. If this is true then the fractions or ratios are equal.

EXAMPLE 1: Verify the proportion of the two ratios, i.e. determine if the proportion is a true statement or not.

$$\frac{25}{100} = \frac{1}{4}$$

cross multiplying

$$\frac{25}{100} \times \frac{1}{4}$$

then verifying

$$4 \times 25 \stackrel{?}{=} 1 \times 100$$

$$100 = 100 \checkmark$$

This proportion is true.

EXAMPLE 2: Verify the proportion of the two ratios, i.e. determine if the proportion is a true statement or not.

$$\frac{33}{70} = \frac{5}{7}$$

cross multiplying

$$\frac{33}{70} \times \frac{5}{7}$$

then verifying

$$7 \times 33 \stackrel{?}{=} 70 \times 5$$

$$231 \neq 350$$

This proportion is not true.

The same is true when we look at proportions involving rates. Proportions with rates, however, have the added complexity of units, since the units do no cancel like they do for ratios. The end result, using cross-multiplication to verify equality, is still the same as it is for ratios, as we now show.

EXAMPLE 3: Verify the proportion of the two rates, i.e. determine if the proportion is a true statement or not.

$$\frac{16\,\text{miles}}{3\,\text{hours}} = \frac{48\,\text{miles}}{9\,\text{hours}}$$

cross multiplying

$$\frac{16\,\text{miles}}{3\,\text{hours}} = \frac{48\,\text{miles}}{9\,\text{hours}}$$

then verifying

$$9 \times 16 \frac{\text{miles}}{\text{hour}} \stackrel{?}{=} 3 \times 48 \frac{\text{miles}}{\text{hour}}$$

$$144 \frac{\text{miles}}{\text{hour}} = 144 \frac{\text{miles}}{\text{hour}} \checkmark$$

This proportion is true. Note, however, that the result of 144 miles/hour is meaningless. Nothing is actually going at this rate. This is just the verification process stating that the two original rates are equal!

Now the alternative is to write this cross-multiplication as:

$$9\,\text{hours} \times 16\,\text{miles} \stackrel{?}{=} 3\,\text{hours} \times 48\,\text{miles}$$

$$144(\text{mile})(\text{hours}) = 144(\text{mile})(\text{hours}) \checkmark$$

However, since the units of (mile)(hours) has no physical meaning we chose the other alternative. The real point is that the numerical coefficients in front of the units are the same. If we want we can ignore the units altogether and simply write:

$$9 \times 16 \stackrel{?}{=} 3 \times 48$$

$$144 = 144 \checkmark$$

This is the approach we will choose to use in the future.

EXAMPLE 4: Verify the proportion of the two rates, i.e. determine if the proportion is a true statement or not. A compressor will compress 30L of air in 15 minutes, or 20L of air in 10 minutes. Are they proportionally equivalent?

$$\frac{30\,\text{L}}{15\,\text{minutes}} = \frac{20\,\text{L}}{10\,\text{minutes}}$$

cross multiplying

$$\frac{30\,\text{L}}{15\,\text{minutes}} = \frac{20\,\text{L}}{10\,\text{minutes}}$$

then verifying

$$10 \times 30 \stackrel{?}{=} 15 \times 20$$

$$300 = 300 \checkmark$$

This proportion is true.

EXERCISES 10.4

Verify the following proportions

1. $\frac{36}{60} = \frac{54}{90}$ Are the two ratios equivalent?
2. $\frac{40}{90} = \frac{8}{18}$ Are the two ratios equivalent?
3. $\frac{15}{75} = \frac{3}{24}$ Are the two ratios equivalent?
4. $\frac{120}{45} = \frac{24}{7}$ Are the two ratios equivalent?
5. $\frac{345}{85} = \frac{21}{5}$ Are the two ratios equivalent?
6. $\frac{12}{16} = \frac{3}{4}$ Are the two ratios equivalent?
7. A car travels 235 miles in 5 hours. Another car travels 188 miles in 4 hours. Are the two rates proportionally equivalent?
8. A car travels 371 miles in 7 hours. Another car travels 212 miles in 4 hours. Are the two rates proportionally equivalent?
9. A car travels 640 miles in 10 hours. Another car travels 330 miles in 6 hours. Are the two rates proportionally equivalent?
10. A climber climbs 2,000 feet in altitude, and the outside air temperature goes down by 42 ^{O}F. Another climber climbs 1600 feet in altitude and the temperature goes down 35 ^{O}F. Are the two rates proportionally equivalent?
11. A climber climbs 1,000 feet in altitude, and the outside air temperature goes down by 25 ^{O}F. Another climber climbs 800 feet in altitude and the temperature goes down 20 ^{O}F. Are the two rates proportionally equivalent?
12. In one glycolysis reaction 3.7mL of CO_2 is produced in 2 hours. In a second glycolysis reaction 8.325mL of CO_2 is produced in 4.5 hours. Are the two rates proportionally equivalent?
13. In one glycolysis reaction 9.75mL of CO_2 is produced in 5.4 hours. In a second glycolysis reaction 2.35mL of CO_2 is produced in 1.7 hours. Are the two rates proportionally equivalent?
14. A compressor compresses 36 liters of air in 21 minutes. A second compressor compresses 48 liters of air in 28 minutes. Are the two rates proportionally equivalent?
15. A compressor compresses 115 liters of air in 60 minutes. A second compressor compresses 175 liters of air in 1.5 hours. Are the two rates proportionally equivalent?

10.5 SOLVING PROPORTIONS

To solve a proportion means we have to verify the proportion containing a variable. We do this by cross-multiplying the proportion as shown above and then finding the value of the variable that makes the proportion true. We first illustrate this process through proportion examples below with ratios with their units removed, and then later we show the same results for proportions involving rates.

EXAMPLE 1: Solve the proportion of ratios: $\frac{x}{3} = \frac{5}{15}$

To solve for the variable x, we can **cross-multiply** the equation by 3 and 15. To **cross-multiply** means we multiply the x by fifteen (15) and the five (5) by three (3). In general, we cross-multiply the denominators times the **opposite** numerators.

This results in the following:

$$15 \times x = 3 \times 15$$

$$15 \times x = 15$$

This tells us that we need to find a number x that when multiplied by 15 gives us 15. We can see that number is simply $x = 1$. That is the value that solves our proportion, or makes it a true statement, i.e.

$$\frac{1}{3} = \frac{5}{15}$$

EXAMPLE 2: Solve the proportion of ratios: $\frac{8}{x} = \frac{7}{5}$

To solve for the variable x, we cross-multiply the equation by the x and the 5.

$$\frac{8}{x} = \frac{7}{5}$$

$$8 \times 5 = 7 \times x$$

$$40 = 7x$$

We now want to find a number that when multiplied by 7 gives us 40. Since no whole number will work, we try the fraction $x = \frac{40}{7}$ as shown below.

$$7 \times \frac{40}{7} = \frac{7 \times 40}{7} = \frac{\cancel{7} \times 40}{\cancel{7}} = 40$$

Notice this is equivalent to simply dividing both sides of $40 = 7x$ by 7. Thus, that will be our approach, to divide both sides of the equation by the coefficient in front of the variable as a last step.

The solution to the proportion is $x = \frac{40}{7}$

$$\frac{8}{\frac{40}{7}} = \frac{8 \div 40}{7} = 8 \times \frac{7}{40} = \frac{56}{40} = \frac{56 \div 8}{40 \div 8} = \frac{7}{5}$$

In the next chapter we explore this process of solving for variables more generally.

In the next example we introduce a very common rate which is speed, $\frac{\text{distance}}{\text{time}}$, and show how to solve proportions related to speed.

EXAMPLE 3: Solve the proportion of rates.

$$\frac{440\,\text{miles}}{8\text{ hours}} = \frac{320\,\text{miles}}{\text{x hours}}$$

Round your answer to the nearest tenths of an hour

To solve for the variable x, we ignore the units, and cross-multiply the equation to obtain:

$$440 \times x = 320 \times 8$$
$$440x = 2{,}560$$

To solve for x, divide both sides of the proportion by 440 and write you answer as a decimal using your calculator. Then round to the nearest tenths place.

$$\frac{\cancel{440}\,x}{\cancel{440}} = \frac{2{,}560}{440}$$
$$x = 5.818\ldots$$
$$x = 5.8$$

This means it would take approximately 5.8 hours.

$$\frac{440\,\text{miles}}{8\text{ hours}} \approx \frac{320\text{ miles}}{5.8\,\text{hours}}$$

EXERCISES 10.5

Solving Proportions

Solve each ratio proportion.

1. $\frac{x}{3}=\frac{4}{5}$

2. $\frac{2}{7}=\frac{b}{6}$

3. $\frac{n}{12}=\frac{3}{8}$

4. $\frac{4}{a}=\frac{3}{10}$

5. $\frac{7}{12}=\frac{5}{d}$

6. $\frac{7}{y}=\frac{9}{8}$

7. $\frac{n}{4}=\frac{3}{8}$

8. $\frac{6}{x}=\frac{3}{2}$

9. $\frac{5}{r}=\frac{2}{7}$

10. $\frac{s}{7}=\frac{3}{14}$

11. $\frac{w}{2}=\frac{9}{16}$

12. $\frac{p}{5}=\frac{8}{9}$

Solve each rate proportion and round your answer to the nearest tenths place as needed.

13. $\frac{200\text{ miles}}{3\text{ hours}}=\frac{250\text{ miles}}{x\text{ hours}}$

14. $\frac{565\text{ miles}}{9\text{ hours}}=\frac{x\text{ miles}}{4\text{ hours}}$

15. $\frac{x\text{ km}}{3\text{ hours}}=\frac{415\text{ km}}{2\text{ hours}}$

16. $\frac{1500\text{ km}}{3\text{ hours}}=\frac{x\text{ km}}{4.5\text{ hours}}$

17. $\frac{x\text{ m}}{5\text{ sec}}=\frac{1575\text{ m}}{3\text{ sec}}$

18. $\frac{4440\text{ m}}{x\text{ sec}}=\frac{532\text{ m}}{8\text{ ecs}}$

10.6 MORE APPLICATIONS OF PROPORTIONS

Proportions can be used to solve a variety of problems across a number of disciplines. In this section we highlight examples from a few areas. We use the cross-multiplication algorithm we introduced in the previous sections to solve proportions, and apply it to some real-life examples.

The examples we consider in this section are called direct proportions. This means that as the value of one objects increases the other value also increases. If, on the other hand, the value of the object decreases the other objects value also decreases. We call this a direct proportion.

We provide a range of problems in this section to highlight the utility of the proportion concept. In the first three examples we choose to set up and solve our proportions as ratios. In the last three examples we solve the same three problems, but instead write them as rates. Both techniques work equally well, and you can choose whichever approach you feel most comfortable with.

EXAMPLE 1: If 2 pencils cost \$1.50, how many pencils can you buy with \$9.00?

Since the cost of the pencils varies directly with their number meaning if we increase the number of pencils the cost also increases, this is a direct proportion.

To solve, we set up the proportion as a ratio by placing like objects above like objects in our proportion, so that our units cancel. We also place related objects directly across from each other in the proportion.

$$\frac{2\,\text{pencils}}{x\,\text{pencils}} = \frac{\$1.50}{\$9.00}$$

To solve for the x pencils, we cancel the units, cross-multiply and solve for x.

$$(2)(9.00) = (1.50)x$$
$$18.00 = 1.50x$$
$$\frac{18.00}{1.50} = \frac{\not{1.50}x}{\not{1.50}x}$$
$$12 = x$$

This means we can buy 12 pencils with $9.00.

EXAMPLE 2: If Dimitri ran 200 meters at a constant speed in 28 seconds. How long did he take to run 1 meter.

Since the distance varies directly with time, this is a direct proportion.

To solve we set up the direct proportion equation:

$$\frac{28\,\text{sec}}{x\,\text{sec}} = \frac{200\text{ m}}{1\text{ m}}$$

To solve for the time, we cancel the units and cross-multiply and solve for x.

$$28 \times 1 = 200 \times x$$
$$28 = 200\,x \; \frac{28}{200} = \frac{\not{200}\,x}{\not{200}}$$
$$0.14 = x$$

It would take Dimitri 0.14 seconds to run 1 meter.

EXAMPLE 3: If $\frac{2}{7}$ of a tank can be filled in 5 minutes, how many minutes will it take to fill the entire tank?

Since the portion of the tank filled varies directly with time this is a direct proportion. To solve we set up the direct proportion equation:

$$\frac{5\,\text{minutes}}{x\,\text{minutes}} = \frac{\frac{2}{7}\text{tank}}{1\text{ tank}}$$

We cancel the units, and cross-multiply

$$1 \times 5 = \frac{2}{7} \times x$$
$$5 = \frac{2}{7}x$$

To solve for x we need to divide both sides of the proportion by the fraction $\frac{2}{7}$. This gives us:

$$5 \div \frac{2}{7} = \frac{2}{7}x \div \frac{2}{7}$$
$$5 \times \frac{7}{2} = \frac{2}{7}x \times \frac{7}{2}$$
$$\frac{35}{2} = \frac{2 \times x \times 7}{7 \times 2}$$
$$17.5 = \frac{\cancel{14}x}{\cancel{14}} = x$$

It would take 17.5 minutes to fill the entire tank.

EXAMPLE 4: If 2 pencils cost \$1.50, how many pencils can you buy with \$9.00?

Since the cost of the pencils varies directly with their number meaning if we increase the number of pencils the cost also increases, this is a direct proportion.

To solve, we set up the proportion as a rate writing the rate as the cost over the number of pencils and equating the two rate. This implies that we have a fixed cost per pencil.

$$\frac{2 \text{ pencils}}{\$1.50} = \frac{x \text{ pencils}}{\$9.00}$$

To solve for the x pencils, we can ignore the units as we did in section 10.4, cross-multiply and solve for x.

$$(2)(9.00) = (1.50)x$$
$$18.00 = 1.50x$$
$$\frac{18.00}{1.50} = \frac{\cancel{1.50}x}{\cancel{1.50}x}$$
$$12 = x$$

This means we can buy 12 pencils with \$9.00.

EXAMPLE 5: If Dimitri ran 200 meters at a constant speed in 28 seconds. How long did he take to run 1 meter.

Since the distance varies directly with time, this is a direct proportion.

To solve we set up the direct proportion equation:

$$\frac{200\text{ m}}{28\text{ sec}} = \frac{1\text{ m}}{x\text{ sec}}$$

To solve for the time, we ignore the units, cross-multiply and solve for x.

$$200 \times x = 28 \times 1$$
$$200\,x = 28$$
$$\frac{\cancel{200}\,x}{\cancel{200}} = \frac{28}{200}$$
$$x = 0.14$$

It would take Dimitri 0.14 seconds to run 1 meter.

EXAMPLE 6: If $\frac{2}{7}$ of a tank can be filled in 5 minutes, how many minutes will it take to fill the entire tank?

Since the portion of the tank filled varies directly with time this is a direct proportion. To solve we set up the direct proportion equation:

$$\frac{\frac{2}{7}\text{ tank}}{5\text{ minutes}} = \frac{1\text{ tank}}{x\text{ minutes}}$$

We ignore the units, and cross-multiply

$$\frac{2}{7} \times x = 5 \times 1$$
$$\frac{2}{7} \times x = 5$$

To solve for x we need to divide both sides of the proportion by the fraction $\frac{2}{7}$. We note that dividing by $\frac{2}{7}$ is equivalent to multiplying by $\frac{7}{2}$. This gives us:

$$\frac{7}{2} \times \frac{2}{7} x = \frac{7}{2} \times 5$$
$$\frac{7 \times 2}{2 \times 7} x = \frac{35}{2}$$
$$\frac{14}{14} x = 17.5$$
$$x = 17.5$$

It would take 17.5 minutes to fill the entire tank.

EXAMPLE 7: Solve the proportion of rates. You traveled 380 miles in 6 hours, how long did it take you to travel 275 miles given that your speed was a constant value. Round your answer to the nearest tenths of an hour

Set up the two rates as miles per hour, and set them equal to each other.

$$\frac{380\text{ miles}}{6\text{hours}} = \frac{275\text{miles}}{x\text{ hours}}$$

To solve for the variable x, we ignore the units, and cross-multiply the equation to obtain:

$$380 \times x = 275 \times 6$$
$$380x = 1{,}650$$

To solve for x, divide both sides of the proportion by 380 and write you answer as a decimal using your calculator. Then round to the nearest tenths place.

$$\frac{\cancel{380}x}{\cancel{380}} = \frac{1{,}650}{380}$$
$$x = 4.3421\ldots$$
$$x = 4.3$$

This means it would take approximately 5.3 hours.

$$\frac{380\text{ miles}}{6\text{hours}} \approx \frac{275\text{miles}}{4.3\text{hours}}$$

EXERCISES 10.6

1. If 5 pencils cost $1.75, how many pencils can you buy with $10.00?
2. If 7 batteries cost $11.50, how many batteries can you buy with $34.50?
3. If 2 pounds of nails cost $2.75, how many many pounds can you buy with $12.00?
4. If 12 pounds of chlorine cost $13.75, how many pounds can you buy with $45.00?
5. If Shirley drove 65 miles per hour constantly for 6 hours. How long did it take her to drive 300 miles.
6. If Juan ran 400 meters at a constant speed in 54 seconds. How long did he take to run 10 meters.
7. If Jessica ran 1 km at a constant speed in 280 seconds. How long did he take to run 20 meters.
8. If Justin drove 58 miles per hour constantly for 4 hours. How long did it take him to drive

150 miles.

9. If $\frac{3}{10}$ of a tank can be filled in 15 minutes, how many minutes will it take to fill the entire tank?

10. If $\frac{3}{8}$ of a tank can be filled in 1 hours, how long will it take to fill the entire tank?

11. If $\frac{5}{9}$ of a tank can be filled in 3 hours, how long will it take to fill the entire tank?

12. If $\frac{7}{12}$ of a tank can be filled in 7 minutes, how many minutes will it take to fill the entire tank?

Chapter 10 Practice Test

Compare the number of objects in the second group to the number in the first, and then explain what the resulting ratio means when written as a fraction. Also rewrite as a decimal approximated to the hundredths place, then as a fraction over a 100, and finally as a percentage.

1. \$125 to \$275
2. 19 homes to 24 homes

Given the following, find the average rates. Be sure to include the appropriate units.

3. A car travels 384 miles in 8 hours, What is its average rate-of-change in miles per hour?
4. As we descend 1,750 feet in altitude, the temperature goes upn by 35 ^{O}F. What is its average rate-of-change of ^{O}F per foot?
5. As you travel 800 meters up a mountain, your temperature went from 26°C to 8°C.
6. A weather balloon rises 75 meters in 10 seconds.
7. In a glycolysis reaction the volume of CO_2 produced is 2.95 mL in 45 minutes.

Verify the following proportions

8. $\frac{12}{20} = \frac{3}{5}$ Are the two ratios equivalent?
9. $\frac{35}{42} = \frac{5}{7}$ Are the two ratios equivalent?
10. A car travels 235 miles in 5 hours. Another car travels 188 miles in 4 hours. Are the two rates proportionally equivalent?
11. A car travels 371 miles in 7 hours. Another car travels 212 miles in 4 hours. Are the two rates proportionally equivalent?

Solve each ratio proportion

12. $\frac{x}{4} = \frac{3}{10}$
13. $\frac{5}{r} = \frac{7}{6}$

Set up and solve each rate proportion

14. If Stanley ran 2,000 meters at a constant speed in 350 seconds, how long did he take to run300 meters?
15. If Rachael drove 648 miles in 5 hours, how long did it take her to drive 175 miles?
16. If $\frac{4}{15}$ of a tank can be filled in 30 minutes, how many minutes will it take to fill the entire tank?

CHAPTER 11

Solving Equations

In chapter 8 we presented some of the basics of the language we call algebra. We introduced the concept of a variable and an expression. In this chapter, we begin to show how to do something useful, i.e. how to solve equations involving expressions expressions.

11.1 SOLVING FOR VARIABLES

To solve for a variable, means we are finding the value or values of the variable that make the equation a true statement. We can think of the equal sign as representing a scale that we must keep in balance. As with a scale we can add, subtract, multiply or divide whatever is on the scale, but to keep it in balance, we must do the same exact thing to both sides of the scale or equation.

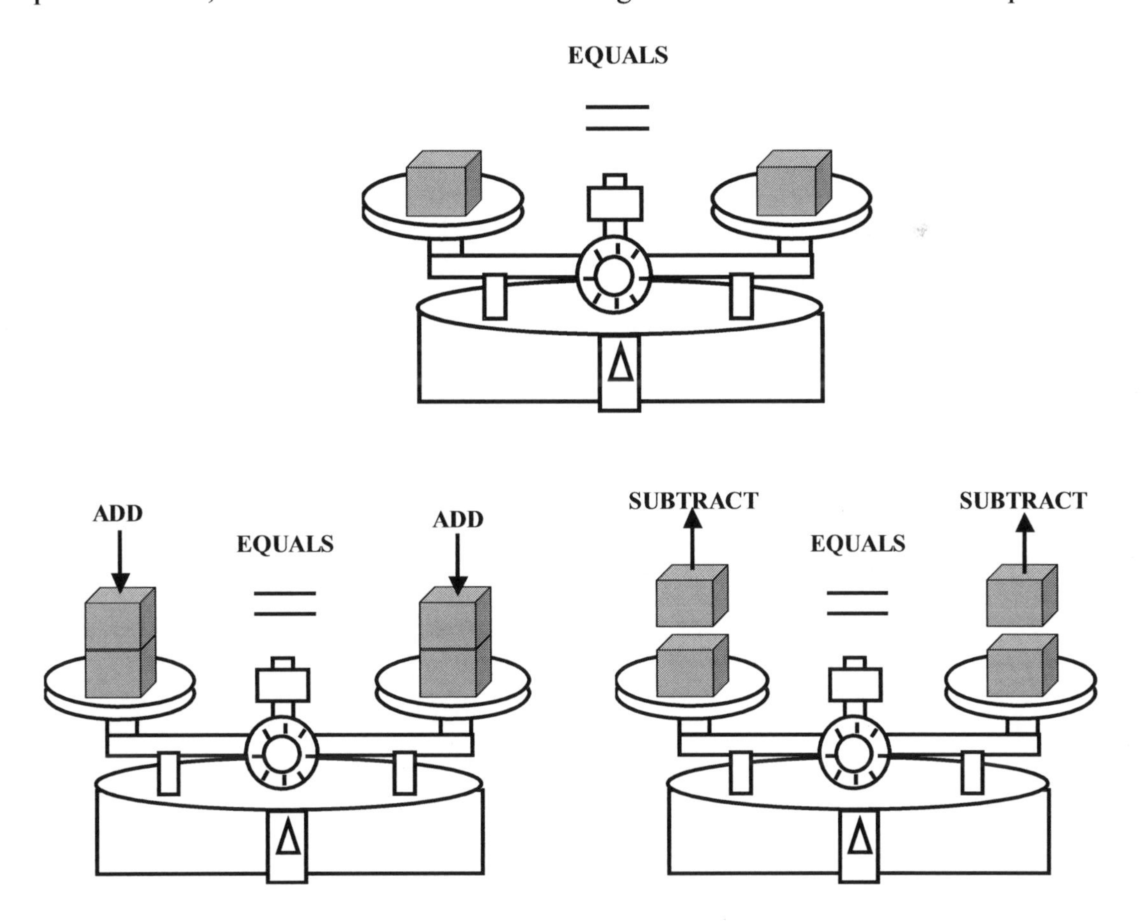

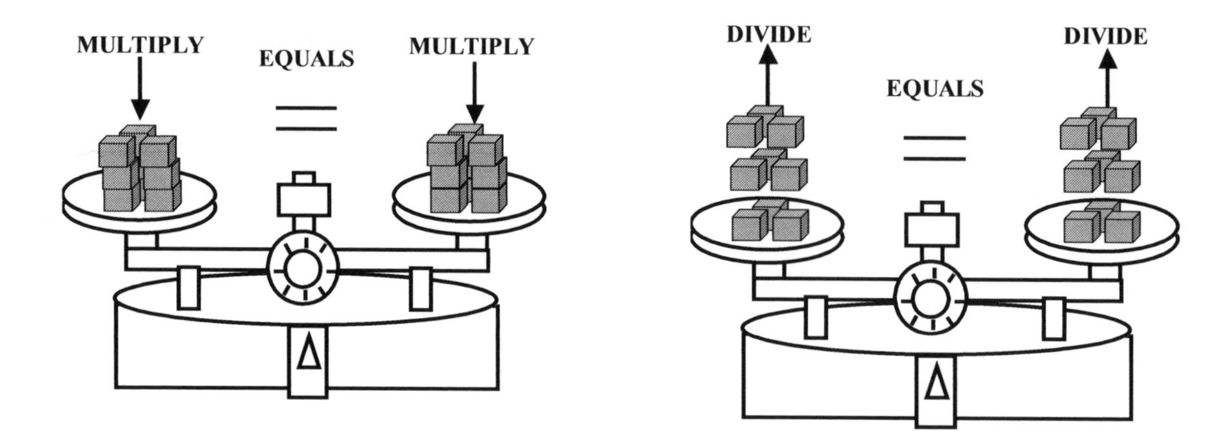

Solving linear equations is a fundamental skill that we'll find useful when solving all the different types of equations, and inequalities in this chapter. Consequently, we will spend time showing how to do this in a simple step-by-step approach.

A linear equation is an equation where two linear expressions are set equal to each other. The objective in solving linear equations is to find the value of the variable that makes the two expressions equal to each other. We accomplish this through a series of arithmetic operations that allow us to isolate the variable, with a unit coefficient, on one side of the equation, and a number on the other.

In solving linear equations in a single variable, we are usually looking for a single solution. There are, however, special cases that will have no solution or will have any value of the variable as a solution. These are more of an exception than the rule, so while we'll show them to you later in this chapter as unique cases, we will not focus on them, and their occurrence can be skipped without any loss of continuity in the material.

11.2 SOLVING ONE-STEP EQUATIONS

These are equations that only require one algebraic step to solve for the variable. Thus, we will either add or subtract a fixed number, or multiply or divide by a fixed number. Consider the following examples:

EXAMPLE 1: Solve the equation for the variable: $x-7=5$

$$\begin{array}{r} x-7=5 \\ \underline{+7 \quad +7} \\ x=12 \end{array}$$

Step 1: Add 7 to both sides

, or

$$\begin{aligned} x-7&=5 \\ x-7+7&=5+7 \quad \text{Step 1: Add 7 to both sides} \\ x&=12 \end{aligned}$$

Notice, we can either add 7 beneath the equation, or inline with the equation. The result is the same.

EXAMPLE 2: Solve the equation for the variable: $a+2=-3$

$$\begin{array}{r} a+2=-3 \\ \underline{-2 \quad -2} \\ a=-5 \end{array} \quad \text{Step 1: Subtract 2 from both sides}$$

, or

$$\begin{aligned} a+2&=-3 \\ a+2-2&=-3-2 \quad \text{Step 1: Subtract 2 from both sides} \\ a&=-5 \end{aligned}$$

EXAMPLE 3: Solve the equation for the variable: $3y=12$

$$\frac{\cancel{3}y}{\cancel{3}}=\frac{12}{3} \quad \text{Step 1: Divide both sides by 3}$$

$$y=4$$

EXAMPLE 4: Solve the equation for the variable: $\frac{p}{-5}=-7$

$$(-5)\frac{p}{-5}=-7(-5) \quad \text{Step 1: Multiply by } -5$$

$$\frac{\cancel{(-5)}}{1}\frac{p}{\cancel{-5}}=-7(-5)$$

$$p=35$$

EXAMPLE 5: Solve the equation for the variable: $-x=4$

This equation is equivalent to $(-1)x=4$, and can be solved in two different ways. The first approach is to multiply both sides of the equation by –1, and the second is to divide both sides of the equation by –1.

Approach 1

$$(-1)(-x)=4(-1) \quad \text{Step 1: Multiply by } -1$$
$$(-1)(-1)x=-4$$
$$x=-4$$

Approach 2

$$\frac{(-1)x}{(-1)}=\frac{4}{(-1)} \quad \text{Step 1: Divide by } -1$$
$$\frac{\cancel{(-1)}x}{\cancel{(-1)}}=\frac{4}{(-1)}$$
$$x=-4$$

Instead of adding or subtracting a fixed number we can also add or subtract a variable from both sides of an equation. Since the variable is the same, we are actually adding or subtracting an equivalent amount from both sides of the equation. This means we are not changing the solution to the equation. Consider the following example:

EXAMPLE 6: Solve the equation for the variable: $2w = w$

$$2w = w$$
$$\underline{-w \quad -w} \quad \text{Step 1: Subtract w from both sides}$$
$$w = 0$$

Unique Cases (Optional): This section can be skipped without losing continuity.

EXAMPLE 7: Solve the equation for the variable

$$x+3=x+5$$
$$x-x+3=x-x+5 \quad \text{Step 1: Subtract x}$$
$$3 \neq 5 \quad \text{Impossible result}$$

No Solution. The equation is a contradiction.

EXAMPLE 8: Solve the equation for the variable

$$w-7=w-7$$
$$w-w-7=w-w-7 \quad \text{Step 1: Subtract w}$$
$$-7=-7 \quad \text{Always true}$$

Any value of w will work, i.e. all Real numbers. The equation is always true (also called a Tautology in the language of logic).

If, when solving an equation, the variable vanishes, the answer is either no solution if you get a contradiction (both sides unequal), or all Real numbers if you obtain a true statement (both sides equal).

EXERCISES 11.2

One-Step Equations

Solve each equation

1. $16 = 7 + x$
2. $5 + a = 9$
3. $w - 3 = 7$
4. $-3 = 16 + y$
5. $p - 8 = 13$
6. $n + 7 = -5$
7. $r - 15 = -7$
8. $5 + t = 31$
9. $-a = 3$
10. $-13 + v = 7$
11. $16x = 48$
12. $-72 = 9w$
13. $-y = -5$
14. $-8 = 4b$
15. $38 = -19a$
16. $-21s = 0$
17. $21 = -7p$
18. $\frac{t}{3} = -4$
19. $2x = 0$
20. $\frac{x}{5} = 10$
21. $17 = -c$
22. $-7 = \frac{m}{4}$
23. $27 = \frac{y}{3}$
24. $a + 11 = 11$
25. $-14 + b = -14$
26. $3y = 2$
27. $72 = -36v$
28. $-6 = \frac{x}{12}$
29. $-5 + k = 28$
30. $-b = -13$
31. $5x = 7$
32. $p - 26 = -28$
33. $x - 26 = x - 28$
34. $6 - y = -y + 6$
35. $3 - w = -w + 4$
36. $13 + z = z - 3$

10.3 SOLVING TWO-STEP EQUATIONS

These are equations that require two algebraic steps to solve for the variable. Thus, we will either add or subtract a number, and then multiply or divide by a number, or vice-versa. Consider the following examples:

EXAMPLE 1: Solve the equation for the variable: $3x + 5 = -7$

$$3x + 5 - 5 = -7 - 5 \quad \text{Step 1: Subtract 5}$$

$$3x = -12$$

$$\frac{\cancel{3}x}{\cancel{3}} = \frac{-12}{3} \quad \text{Step 2: Divide by 3}$$

$$x = -4$$

EXAMPLE 2: Solve the equation for the variable: $5u-6=15$

$$5u-6=15 \quad \text{Step 1: Add 6}$$
$$5u-6+6=15+6$$
$$5u=21$$
$$\frac{\not{5}u}{\not{5}}=\frac{21}{5} \quad \text{Step 2: Divide by 5}$$
$$u=\frac{21}{5}$$

EXAMPLE 3: Solve the equation for the variable: $2w+3=w-5$

$$2w+3-3=w-5\ -3 \quad \text{Step 1: Subtract 3}$$
$$2w=w-8$$
$$2w-w=w-8-w \quad \text{Step 2: Subtract w}$$
$$w=-8$$

EXAMPLE 4: Solve the equation for the variable: $6s=3s$

$$6s-3s=3s-3s \quad \text{Step 1: Subtract 3s}$$
$$3s=0$$
$$\frac{\not{3}s}{\not{3}}=\frac{0}{3} \quad \text{Step 2: Divide by 3}$$
$$s=0$$

If we think about this last example further we can see that it makes sense. We are asking, what number when multiplied by 6 equals the same number multiplied by 3? The only possible number for which this is true is zero. Any other number would give a different result.
We should also note that while we might be tempted to divide both sides of this equation by the variable, s, that this would cause problems, since

$$6s=3s$$
$$\frac{6\not{s}}{\not{s}}=\frac{3\not{s}}{\not{s}}$$
$$6=3$$

In this way we get a contradiction and we might be tempted to say that there is no solution to this equation. However, the reason we get a contradiction is that the actual solution to this equation is s = 0. Thus, when we divided by s, we are actually dividing by zero. We should recall that we are never allowed to divide by zero. Since we performed an illegal operation, we got an incorrect result. We violated the rules of the game, so we can't expect the answer we get to be correct.
Multiplying or dividing by variables is one area where we have to be cautious. We always have to consider the possibility that the variable is zero!

EXERCISES 11.3

Two-Step Equations

Solve each equation

Basic Problems

1. $4 = 2a - 2$
2. $10 - 3x = -5$
3. $-16 + 5x = -9$
4. $9 = -3 + 4p$
5. $2x + 5 = 18$
6. $4p - 3 = -8$
7. $16 + u = 16$
8. $7 + 7c = 7$
9. $-10z + 1 = -99$
10. $5 - 3a = 12$
11. $13 - 3w = -7$
12. $1 - 4x = 17$
13. $2b - 13 = 13$
14. $7a + 24 = -4$
15. $-20 = 5p + 3$
16. $5x + 3 = 7$
17. $4c - 16 = 9$
18. $7 = 2x - 6$
19. $2p + 3 = 16$
20. $-7b - 5 = 21$

Intermediate and Advanced Problems

21. $-9 = \frac{n}{3} + 11$
22. $\frac{s}{21} - 3 = -4$
23. $13 + \frac{u}{-2} = 11$
24. $17 + \frac{y}{-5} = -3$
25. $\frac{p}{11} - 3 = 2$
26. $-5 = 5 + \frac{r}{4}$
27. $\frac{u}{-5} + 7 = -9$
28. $7 + \frac{x}{-5} = 3$
29. $5z + 3 = 4z - 2$
30. $-6 = \frac{v}{3} - 7$
31. $\frac{z}{7} + 5 = 7$
32. $7r = 2r$

11.4 SOLVING MULTI-STEP EQUATIONS

EXAMPLE 1: Solve the equation for the variable: $2a - 3 + 5a + 1 = a - 7$

$$2a - 3 + 5a + 1 = a - 7 \quad \text{Step 1: Combine like terms}$$

$$7a - 2 = a - 7$$

$$7a - 2 + 2 = a - 7 + 2 \quad \text{Step 2: Add 2}$$

$$7a = a - 5$$

$$7a - a = a - a - 5 \quad \text{Step 3: Subtract a}$$
$$6a = -5$$
$$\frac{\cancel{6}a}{\cancel{6}} = \frac{-5}{6} \quad \text{Step 4: Divide by 6}$$
$$a = \frac{-5}{6}$$

EXAMPLE 2 Solve the equation for the variable: $1-(2-3v)=4(v-7)-6$

$$1-2+3v = 4v-28-6 \quad \text{Step 1: Use Distributive Property}$$
$$-1+3v = 4v-34 \quad \text{Step 2: Combine like terms}$$
$$-1+34+3v = 4v-34+34 \quad \text{Step 3: Add 34}$$
$$33+3v = 4v$$
$$33+3v-3v = 4v-3v \quad \text{Step 4: Subtract 3v}$$
$$33 = v$$

EXERCISES 11.4

Solving Multi-Step Equations

Solve each equation

Basic Problems

1. $8=-8x+2x-4$
2. $3a-4=2a-3$
3. $1-2y=3y-9$
4. $4c-3=2c-11$
5. $p-3=p+5$
6. $3-z=7-z$
7. $9=-3(-x+4)$
8. $-(2x+5)=12$
9. $2(4y-3)-7=y-5$
10. $24=-8(2s+3)$
11. $-(2n-3)=18$
12. $2r+3(r+1)=-5$
13. $6(2m-5)=-18$

Intermediate and Advanced Problems

14. $2(p+2)-4(2p-1)=18$
15. $2n-6=-3(n+5)$
16. $4-3w=6(2w+5)$
17. $-6(u+1)+4(u-1)=3(u+2)$
18. $-(a+3)+4(1-a)=7(2a+1)$
19. $5-3(1-5a)=1-3a$
20. $16k-12=-(2-3k)$
21. $13-21t=3(4-2t)$

22. $7(x+4)-3(1-2x)=-36$

23. $32=-5(y+4)+3(y-2)$

24. $-(3-5p)=6-2p$

25. $16-3v=2-3(v+5)$

26. $-x+17+3x=5-2(1-2x)$

11.5 APPLICATIONS OF EQUATIONS

In this section we demonstrate how to use linear equations to solve application problems. The problems themselves are fairly simple and they are mainly meant as an aid for learning how to set up and solve some beginning problems. Thus, we will focus more on the approach to setting up and solving the problem.

The first step is to read the problem thoroughly, a few times, not just once. Write down what you are asked to find. Determine what your variable (unknown you are solving for) is. How is it related to itself or other numbers? This last question gives us the equation that we have to solve. Write down the equation, and reread the problem when looking at the equation to be sure that you have not missed or confused anything. Does the equation you have written represent what is being asked? In this course the equation will either be a linear or a quadratic equation. Solve the resulting equation. Does the result make sense, or can it be tested?

We illustrate the approach through examples:

EXAMPLE 1: Twelve more than six times a number is thirty-six. What is the number?

First, we notice that we are looking for a number, so we use "n" to represent the unknown number. Next, we convert the sentence above into an equation. We see that twelve more than six times a number is 36 is equivalent to, six times a number plus twelve equals thirty-six, or

$$6n+12=36$$

Solving this equation using the techniques developed above we find that $n=4$.

Does it work? Since, $6(4)+12=24+12=36$, we see that we have found the correct solution.

EXAMPLE 2: Five less than four times a number is twenty-seven. Find the number.

Again we use "n" to represent the unknown number. Next, we convert the sentence above into an equation. We see that five less than four times the number is equivalent to $4n-5$. Setting this equal to 27 we have:

$$4n-5=27$$

Solving this equation using the techniques developed earlier we find that $n=8$.

EXAMPLE 3: The length of one side of an equilateral triangle is 6 inches. What is the perimeter of the triangle?

First, we use "p" to represent the unknown number and ℓ to represent the length of its side. Next we use the fact that an equilateral triangle has three equal sides. Thus, we are trying to find $p = 3\ell$, where ℓ is the length of one side of the triangle. Next, we substitute the know values into the perimeter equation to obtain:

$$\begin{aligned} p &= 3\ell \\ &= 3(6 \text{ inches}) \\ &= 18 \text{ inches} \end{aligned}$$

EXAMPLE 4: A triangle has three interior angles such that one angle is twice the value of the first angle and the last angle is five times the size of the first angle. What are the values of all the angles?

Let "a" represent the value of the first angle. The values of the other two angles are $2a$ and $5a$ respectively. Next we use the fact that the sum of the interior angles of a triangle equals 180 degrees. Thus, we are trying to find "a" such that:

$$a + 2a + 5a = 180^\circ$$

Solving this we obtain

$$\begin{aligned} 8a &= 180^\circ \\ a &= \frac{180^\circ}{8} \\ a &= 22.5^\circ \end{aligned}$$

The first angle is 22.5°, the values of the other two angles are: $(2)22.5^\circ = 45^\circ$, and $(5)22.5^\circ = 112.5^\circ$ respectively. Notice that they do sum to 180°.

EXAMPLE 5: The perimeter of a rectangle is 360 feet. Its length is twice its width. What are the length and width of the rectangle?

The formula for the perimeter of a rectangle is: $p = 2\ell + 2w$

With w equal to the width of the rectangle, the length is given by: $\ell = 2w$. Replace ℓ in the perimeter equation with 2w, and the perimeter with 360 to obtain:

$$\begin{aligned} p &= 2\ell + 2w \\ 360 &= 2(2w) + 2w \\ 360 &= 4w + 2w \\ 360 &= 6w \\ \frac{360}{6} &= \frac{\cancel{6}w}{\cancel{6}} \\ 60 &= w \end{aligned}$$

Thus, w = 60 feet and ℓ = 2(60) feet = 120 feet.

EXERCISES 11.5

Applications of equations

1. Six more than three times a number equals thirty-three. Find the number.
2. Five less than six times a number is seven. What is the number?
3. Twenty-five more than eight times a number is one. Find the number.
4. Seven less than two times a number is negative fifteen. Find the number.
5. The perimeter of a rectangle is sixty-four inches. Its length is three times as long as its width. What are the length and width of the rectangle?
6. The perimeter of a rectangle is 150 feet long. Its length is twice its width. What is the length and width of the rectangle?
7. The perimeter of an equilateral triangle is thirty-six inches. What is the length of each side?
8. The perimeter of an equilateral triangle is twelve feet. What is the length of each side?
9. If the interior angles of a triangle are such, that the largest angle is three times as large as the smallest angle, and the middle angle is twice the smallest angle, what are the angles of the triangle?
10. If the interior angles of a triangle are such that the largest angle is five times as large as the smallest angle, and the middle angle is three times the smallest angle, what are the angles of the triangle?
11. The sum of two consecutive integers is twenty-three. What are the numbers?
12. The sum of two consecutive integers is thirty-nine. What are the numbers?
13. One number is twice another number, and their sum is thirty. What are the numbers?
14. One number is three times as large as another number. Their sum is eight. What are the two numbers?
15. You are given three numbers, where the largest number is three times the smallest number, and the middle number is two times as large as the smallest number. The sum of all three numbers is thirty-six. What are the three numbers?

Chapter 11: Practice Test

Solve each linear equation

1. $z-3=-3$
2. $5y-2=-17$
3. $w-3=7$
4. $p+8=2$
5. $-6=\frac{x}{12}$
6. $10-3x=-5$
7. $9=-3+4p$
8. $3a-4=2a-3$
9. $16-3v=2-3(v+5)$

10. The perimeter of a rectangle is 80 inches. Its length is 3 times as long as its width. What are the length and width of the rectangle?

11. The sum of two consecutive integers is forty-seven. What are the numbers?

84118901R00133

Made in the USA
San Bernardino, CA
03 August 2018